Penoyre & Prasad

Retrofit for Purpose

Low Energy Renewal of Non-Domestic Buildings

RIBA ₩ **Publishing**

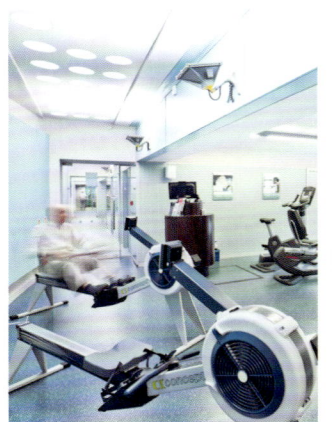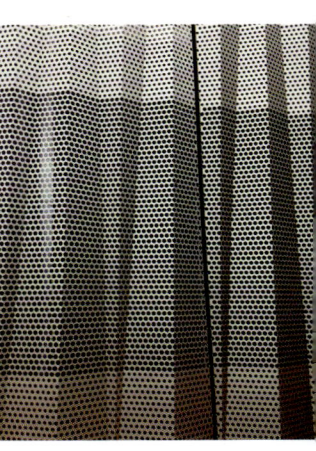

Published by RIBA Publishing,
15 Bonhill Street, London EC2P 2EA

ISBN 978 1 85946 514 1

Stock code 80589

Designed and typeset by: Alex Lazarou
Printed and bound by: Butler, Tanner & Dennis
Frome & London, UK

RIBA Publishing is part of RIBA Enterprises Ltd.
www.ribaenterprises.com

Acknowledgements

With grateful thanks to:

- all the architects, clients and engineers who provided information for the case studies
- all the expert contributors for their essays
- Sarah Drake of Penoyre & Prasad who managed the images and graphic quality
- Hadas Keren and Phyllida Mills of Penoyre & Prasad who helped with researching the case studies
- Matthew Thompson who helped with the case study texts
- Sharla Plant, Kate Mackillop, James Hutchinson and the team at RIBA Publications
- and, finally, James Thompson who prompted us to produce this book.

Sunand Prasad
February 2014

Contents

SUNAND PRASAD

RETROFIT IN PRACTICE

INTRODUCTION

This is a book about altering, remodelling, recladding and refurbishing buildings to achieve reductions in operational energy consumption. It is not a technical guide – there being a steady stream of such guidance readily available – but a combination of case studies and essays that explore the wider context of energy-conserving retrofit projects, addressing the questions: How can they be financed? How does government policy impact on them? How do we tell whether the intended energy savings have been achieved? How can the science of measuring performance inform design? What can we learn from other countries where low energy retrofit is gaining momentum?

FACING PAGE Guy's Hospital Tower overclad, London. See Case Study 11

The competition brief for the recladding of Guy's Hospital Tower, issued in 2008, 35 years after the Tower was first occupied, required that the building's envelope had to be 'fit for purpose' for the next 30 years. It had become unfit because some of its concrete panels were beginning to disintegrate, posing potential danger, its windows were rusting and its facades generally were looking dilapidated and dowdy – a poor reflection on the status and public image of one of the largest NHS Trusts. The huge waste of energy through the old, uninsulated facade was not itself a driver of the decision to reclad; however, Guy's and St Thomas' NHS Foundation Trust was committed to reducing its carbon emissions and asked that solutions should be sustainable and energy efficient.

'Fit for purpose' sounds categorical, but depends heavily on exactly what 'purpose' is intended.[1] When built, Guy's Tower, with its adventurous engineering and its design strategy enshrining flexibility of use, was considered to be outstandingly fit for purpose. In use, the Tower has largely validated this aspect of its design: its floorplates having proved capable of converting easily to make laboratories and offices as well as accommodating changing hospital practice. But the Tower also belongs to a period of architecture when old crafts-based knowledge of what gives building longevity had been replaced by immature theories of building science, prematurely put into large-scale practice. Judging its quality against the timeless Vitruvian ideals of Commodity, Firmness and Delight, it was doing well on Commodity and poorly on Firmness; parts of it literally crumbling away, let alone the energy wasted through its walls. As for Delight, its occupants loved the views over London but few found delight in the Tower's appearance.

'Fit for purpose' for the next 30 years, from the point when the Guy's Tower recladding project will be completed at the beginning of 2014, takes us to 2044, six years short of the deadline by which global carbon emissions must be halved in order to avoid catastrophic global warming, with most of the western world's emissions having to be reduced by between 80 and 90%. By then, the whole of our existing building stock will have to be consuming

radically less energy in use. Therefore, on energy performance grounds alone, almost all of the current stock will be unfit for purpose.

CONSERVING ENERGY: ENERGISING ARCHITECTURE

Although this is a book about retrofitting buildings to save energy, few projects have that aim as the only goal. If the return on investment in the extensive work required to radically reduce energy demand relied on energy savings alone, few clients would undertake the project. The wonderful thing about the buildings illustrated here is that their architectural design, whether modest or assertive, is closely integrated with their energy conservation. The huge value added by architecture made the business case for the project. At Elizabeth II Court, Winchester, for example, clever design enables more people to work in better conditions within a smaller overall space, thus directly saving occupancy costs for the client. Furthermore, the client recognised that improving the work environment helps to increase productivity and reduce staff turnover. At 199 Bishopsgate, the retrofit project was driven by the client's corporate commitment to sustainability and high-quality design as part of their very effective positioning in the market. At Westborough School, the architectural design makes manifest the energy-conserving aspects so as to make the design and construction of the building an integral part of the students' learning about the environment. There are parallel stories behind all these projects and the images speak for themselves.

It is a very good thing that clients have wider reasons for refurbishing buildings than simply enhancing energy efficiency: improving the usability, the appearance, the enjoyment, the meaning of a building for its users through architecture; improving the functionality, performance and spatial quality of the building and perhaps enhancing its image and relationship to its surroundings. Almost all buildings need periodic renewal and over the next decades we have to use the renewal cycle of the buildings as a lever to accomplish the otherwise formidable challenge of energy-efficient retrofit.

THE CARBON EMERGENCY AND EXISTING BUILDINGS

Achieving large reductions in energy consumption is one of the biggest and most urgent challenges for the built environment professions and industries today. The argument for the necessity of reducing the energy consumption of existing buildings is well rehearsed. It has a number of powerful components: the imperative to reduce carbon emissions, the (in)security of energy supply, the rising price of energy and the consequent increase in fuel poverty, to name the principal ones. We know that a significant portion of the reduction can be achieved through modification of behaviour at both individual and corporate/organisational levels. However, the necessary reduction in carbon emissions cannot be achieved unless we ensure that a building's fabric can maintain an acceptable level of comfort in the full range of prevailing local climatic conditions for the smallest possible expenditure of energy.

By 2020, UK Building Regulations are likely to ensure that new buildings – both homes and non-domestic buildings – are generally designed to a standard of energy efficiency that approaches the optimal. In theory, together with changes in behaviour, the decarbonisation of the electricity supply, and the eventual shutdown of the use of gas and oil as heating fuels, new buildings will be designed to 'do their bit' for the post-fossil fuel future, which is the only future with a chance of avoiding catastrophic climate change. However, 'designed to' is not the same as 'constructed to' or 'operated to' achieve that same standard. There is significant evidence now of the 'performance gap': that buildings in use are typically more than twice as energy-hungry as they have been designed to be. This gap casts its shadow, if gaps can cast shadows, over many of the essays in this book. But, even taking that into account, we can at least see that considerable advances have been made in terms of the energy efficiency of new buildings.

However, new buildings yet to be constructed will form only about one-fifth of the total stock of buildings that will be standing in 2050. The vast majority, four-fifths of the total, will have been constructed before the urgent need to curtail carbon emissions was understood.

It is over 40 years since concern about our energy consumption came fully to public attention. In 1972, RIBA President Alex Gordon coined the slogan 'Long life, loose fit, low energy' to describe the ideal building, two years before the world oil crisis drove the issue of energy up the political agenda. It was not until the 1980s that thermal efficiency – or 'conservation of power' – entered the Building Regulations and started the steady, though slow, improvement of the energy performance of new buildings. In the early 1990s, the evidence that the planet was warming in an unprecedented way, and that the warming was caused by the burning of fossil fuels, began to be taken seriously by governments. The turning point was the political consensus that eventually coalesced in the early 2000s, that climate change was an unprecedented threat to civilisation and required concerted international action. The evidence has steadily mounted, the latest example being the Fifth Assessment Report of the Intergovernmental Panel on Climate Change.[2] Although the political consensus has degraded, the momentum that has gathered generally in society has ensured a steady growth in the low carbon economy, assessed at £120 billion of sales annually in the UK.[3] The property and construction industry form a major part of this sector, and it is the norm to assess the performance of buildings in terms of sustainability – within which energy use and carbon emissions are particularly significant elements. Notably, despite the 2008–12 recession, the UK Government has adhered to the decision that all homes will be 'zero carbon' by 2016. That does not mean that new homes will necessarily create zero-carbon emissions, as the original 'zero-carbon' standard has been altered to exclude unregulated emissions (i.e. 'plug loads') – emissions resulting from householders' expenditure of energy through the use of appliances. It is expected that regulations for non-domestic buildings will achieve a similar impact by 2020.

However, Building Regulations barely touch the existing stock. Even when buildings are being converted, the statutory mechanisms generally affect only the parts that are being altered. And yet, as is pointed out in every seminar and conference on climate change, around 80% of the building stock that we will be inhabiting in Europe in 2050 already exists. The year 2050 is the date by which the science tells us we have to be emitting only 20% of the current annual tonnage of greenhouse gases (GHG).[4]

This figure of an 80% reduction in GHG emissions is enshrined in British law through the Climate Change Act, which requires government to set five-yearly carbon budgets, starting in 2008. The carbon budget for the current period (2013–17) is 2,782 $MtCO_2e$, an average of about 560 $MtCO_2e$ per year.[5] Currently, emissions from the operation of non-domestic buildings are running at 48 $MtCO_2e$ per year, or 11.5% of the total. As Figure 1.1 shows, emissions from non-domestic buildings amount to one-quarter of all buildings-related emissions.

However, currently we do not have effective regulatory mechanisms, any evident financial means, or the political will to make the existing building stock sufficiently energy efficient to achieve this target. At the time of writing, the rules for applying value added tax discriminate against the retrofitting of existing homes, subjecting the process to tax at 20%, whereas new build attracts no VAT.

Apart from the scale of the opportunity for carbon emission reduction, there is another powerful reason for retrofitting existing buildings, particularly where that is an alternative to building anew. Figure 1.1 shows that annual UK carbon emissions from construction actively related to non-domestic buildings currently amount to 9% of all built environment related emissions, all but a fraction of which will be from new buildings construction. The energy expended in construction – 'embodied' or 'embedded' energy, to distinguish it from operational energy[6] – has typically amounted to between 13 and 15% of the whole-life use of energy of modern buildings. Retrofitting a building uses much less energy in construction compared with a new building of the same size. As we make buildings more and more energy efficient, the embodied component becomes more significant. Extending the life of a building, so

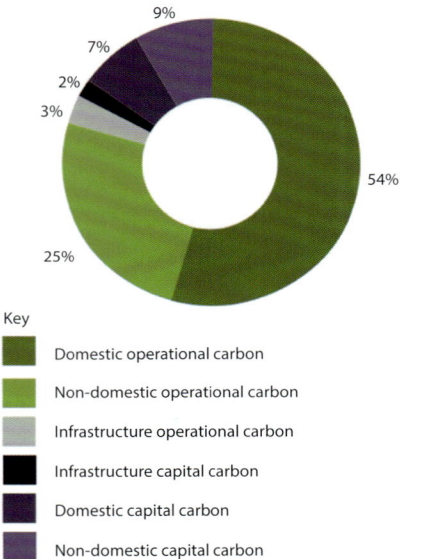

1.1 Breakdown of annual carbon emissions by sector (2010). Operational carbon means emissions in use. Capital carbon means carbon emissions in manufacture, transport and construction

spreading the energy investment in construction over a longer period, is intrinsically energy efficient and consequently limits carbon emissions.

Making the bold and optimistic assumption that the world will eventually take the steps necessary to limit climate change, design, construction and property industries everywhere are going to be very busy over several decades transforming the built environment. So there is a mountain to climb and we are not even in the foothills. Schools of engineering and architecture focus almost exclusively on designing new buildings and calculating their performance. Refurbishment, let alone low energy retrofit, does not have the status even of a sub-discipline. There is no shortage of knowledge of the arts and science of energy-efficiency retrofit; after all, the building physics principles are the same as those for new buildings and refurbishment is a common activity. However, this knowledge is only now coming together as a coherent body of theory and practice to create conditions for real innovation.

This book is a contribution to assembling such a body of knowledge. Although it draws on work and research in academia, this is a practical rather than an academic book. It is intended to help building owners, designers, builders and managers to take a holistic view of their retrofit projects and to learn from others who have completed retrofit projects.

While there are many technical similarities between domestic and non-domestic retrofit, there are crucial differences, particularly in finance, procurement and logistics. There are also technical differences, most evidently in terms of scale, the range of available technologies and the functional requirements. This book focuses on non-domestic retrofit, in the knowledge that others are producing works targeted at the domestic sector and that domestic retrofit has received more attention through programmes such as the Technology Strategy Board's 'Retrofit for the Future' and the UK Government's 'Warm Front' and 'Green Deal' programmes.

CULTURE, FINANCE AND TECHNOLOGY

In the drive to mitigate climate change through reducing anthropogenic carbon emissions we encounter a complex relationship between people's behaviour, available technology and money. In the built environment, the use of which is responsible for half of UK carbon emissions,[7] there are three clear ways to reduce emissions: reduce demand for energy; increase efficiency in the use of energy; and reduce the carbon intensity of the energy that is needed. As Bill Bordass has said: halve the demand, double the efficiency, halve the CO_2 per unit of energy and we will have reduced emissions to one-eighth.

Reducing energy demand

Reducing demand is almost entirely a matter of human behaviour. For example, what temperature is considered to be comfortably warm or cool; whether conscientiously turning off lights is seen as obsessively penny-pinching or admirably disciplined; whether there is an emotional or moral view about waste; whether simplicity is considered attractive or just boring. Attitudes that underpin these behaviours are culturally formed and bound up with the question of individual rights and collective responsibility.

Increasing efficiency in energy use

Increasing efficiency in the use of energy appears initially to be simply a matter of design, technology and finance. However, though they are fundamental, design, technology and finance are conditioned by culture, since the choices that a designer and client make about where to invest available resources are subjective and depend on their priorities and value systems. The regulatory environment in which the design and construction is carried out is also culturally conditioned. A government may decide that regulations are a brake on business and growth and therefore refuse to employ them to the extent that other governments might be prepared to in order to achieve energy efficiency. The UK Department for Business, Innovations and Skills' report[8] on achieving a low carbon built environment found that the biggest obstacle to reducing emissions from the built environment was available finance, not available technology. There were simply no mechanisms in place to release fast enough the huge investment needed to retrofit the existing stock to the levels implied by the carbon budgets set by the Government. However, the bank bailout that followed the 2007/08 credit crunch cost more than retrofitting the entire UK building stock.

Good technological intentions can also flounder upon the rocks of culturally entrenched behaviour. Complex automated environmental systems designed and installed with the aim of attuning energy use more precisely to weather and comfort needs have repeatedly proved to be hugely disappointing in use, as elaborated in essays here by Bill Bordass and Roderic Bunn. It is natural for people to interact with their environment – to take action when they are too hot or too cold, such as opening a window or turning up a thermostat. Systems that deny such behaviours will tend to be fragile.

Low carbon energy

The challenge of reducing the carbon intensity of energy has many similarities to the quest to increase efficiency through design. It requires significant investment to bring about innovation in low carbon energy generation, but the scale of this investment is within normal parameters for the energy sector. The issue is how to shift the investment towards renewable energy sources and away from fossil fuels, on which depend the fortunes of the world's energy companies. That is why carbon capture and storage is seen by many as the only solution, despite the fact that the technology is not developing as fast as was hoped.

Rethinking the designer's remit

The implication of all this for retrofit is that to truly achieve the energy reduction that global warming calls for, and clients increasingly want, designers have to enter into territory that is outside their conventional realm of operation. The essays in this book deal primarily with this unfamiliar territory – with behaviour, with measurement and finance – rather than design and technology, on which there is plentiful, easily-sourced professional knowledge, as indicated in the Further Reading section.

While we have focused in this book on the imperative of reducing energy use, other key indicators of sustainability, in particular the conservation of water and other resources, are no less important. The knowledge base for these areas is well developed and transfers easily from new buildings to retrofit. Equally, the retrofit project presents an opportunity to adapt buildings to the climate change that is now known to be inevitable. The knowledge base for climate adaptation measures for buildings is growing rapidly and applies equally to new buildings and retrofit.[9]

RETROFIT IN PRACTICE

Setting aside the domestic building stock (about 26 million homes), the remaining 3.5 million buildings in the UK present a potentially baffling heterogeneity of type, age, context, tenure, condition, use, build and design quality and historical importance. Imagine a matrix with various types of buildings as the headings running along the top and a set of age ranges down the side – let us say an 8 x 8 matrix. Imagine adding another eight variables of physical context to make a third dimension; then a set of such matrices for tenure and, again, a further set to cover condition. All these variables quickly multiply to thousands of possible kinds of retrofit project. Add to this the different purpose each client will have for commissioning the retrofit project, and the different budgets each will set, and the range of possibilities becomes practically infinite. None of this is surprising, as these buildings would have been built for thousands of different purposes in different circumstances in the first place.

While such heterogeneity of problems and solutions is intrinsic to architecture and construction, its significance for retrofit stems from the inconvenient truth that, for the foreseeable future, retrofit projects are not going to achieve anything like the energy savings that ultimately will be required in each building being retrofitted. For each retrofit project to make a proportionate contribution to the UK's (or anywhere else's) overall target of reducing carbon emissions requires a 'deep retrofit'.

DEEP RETROFIT AND PARTIAL RETROFIT

To qualify as a deep retrofit – shorthand for 'deep energy retrofit' – the entire fabric and the environmental systems of the building have to be evaluated, redesigned and reconstructed in an integrated way. As some of the case studies in this book show, through deep retrofit the performance of existing buildings can equal that of new ones. While deep retrofit is a widely used and understood term, there is no clear terminology for other retrofits. We will use 'partial retrofit' to describe projects of alteration or refit of elements that affect energy use of buildings. Such retrofit may involve only the fabric, such as the cladding on Guy's Tower (see the case study later in this book) or only the active environmental systems, as was primarily the case with 199 Bishopsgate (also presented later as a case study). While the retrofit may be partial, the architectural component of the retrofit project may make a dramatic difference to the quality of the building and the experience of the users.

The budgets available for retrofit today are a fraction of what is required for deep retrofit of the entire building stock, even if undertaken over several decades. One way of looking at this issue is that present costs of energy and the costs of retrofit are such that the payback period for deep retrofit is of the order of 25 years plus, rather than the 5–10 years that would make the investment attractive to most clients. There are good reasons, other than financial ones, for not yet undertaking deep retrofit on a large scale: some of the technologies that achieve operational energy reduction beyond 40–50% are as yet immature or relatively expensive, and likely to become much cheaper over time. New technologies will also emerge as the market in low energy retrofit achieves scale and momentum.

Because deep retrofit is not going to be an option for the majority of projects, choices have to be made as to where retrofit measures should be focused. A key issue to consider is how to avoid compromising the implementation of a higher level of retrofit in the longer term. Measures implemented now could soon become obsolete, and even become obstacles. Insulation, draught seals and plant may need to be stripped out at some future point, wasting both money and the carbon embodied in the materials. That is why projects in their myriad circumstances have to be considered individually in their totality, and one cannot, at least for the time being, draw up sets of standard solutions or a complete retrofit 'kit of parts' for non-domestic projects. Such package-based approaches will emerge as the industry accumulates more of the kind of experience that is documented in the case studies here.

Advocates of a general policy of deep retrofit say that it is better to ensure that a smaller number of buildings are given a deep retrofit – a single shot of medicine that will see them through to the post-fossil fuel times – rather than a larger number fitted with a variety of easier, lighter touch, measures. There is some research[10] to indicate that the cyclical nature of building refurbishment generates a 'lock-in' effect. A building is likely to be refurbished only once every 30 years or so, and consequently buildings that are insufficiently deeply retrofitted will not get a second chance to be adequately retrofitted within governmental carbon reduction timeframes. However, it is difficult to see, outside of a command economy or an unfeasible funding programme, how any government could enforce deep retrofit. In any case, as has been argued above, there are good reasons to let lighter touch retrofit flourish, provided it does not significantly compromise later deeper retrofit.

The way to square the circle and avoid compromising subsequent deeper measures is to see the retrofit project as the first step in a larger, longer term plan to prepare the building for a zero or low carbon future. We can think of such an approach as the equivalent of drawing up a master plan. Like any good master plan, the retrofit plan should have a clear structure while remaining flexible to avoid compromising future possibilities. For example, if the walls are being insulated internally with finishes being ripped out, it is likely to be a false economy not to make to make the building's envelope more airtight or not to design out cold bridges. The work to any element should be fit for the long term.

PURPOSE

Whether or not energy efficiency is at the heart of each refurbishment project's brief, architects and other designers could think of every refurbishment project as an opportunity to start a *parallel project* of optimising the energy performance of the building. Thinking of energy performance as a parallel project helps to position it as a set of activities going beyond design and construction and lasting longer than the refurbishment project itself. The most effective low energy retrofit projects will include modifying user behaviour, a major factor in the actual energy use of a building. They will also integrate the operation of the building and the critical role of the manager/caretaker into the project. Other essays in this book, such as those by Roderic Bunn and Rajat Gupta and Matt Gregg, clearly set out how design and construction alone cannot achieve the energy efficiencies that are required and that clients increasingly seek. In practice, the client may paradoxically be unwilling to include elements that are not clearly identifiable as falling within the domain of design and construction as part of the designer's or project manager's commission; however, we still think there is value in setting out the full picture in this way to encourage take-up of these elements in the future.

RETROFIT VARIABLES

Type

Hours of operation, numbers of occupants, types of activity, intensity of equipment use and a number of other such factors mean that different kinds of buildings exhibit different patterns of energy consumption, and therefore require different design approaches. In addition, the obvious impact of these elements on the planning and design of retrofit projects make targets and benchmarks for energy use dependent on the type of retrofit project under consideration. The availability of benchmark data for retrofit projects is patchy but best practice targets for new buildings are available from CIBSE, the Chartered Institute of Building Services Engineers.[11] There is some sector-specific benchmarking guidance, particularly for schools, accessible via the web. Perhaps the most promising is Carbon Buzz (www.carbonbuzz.org) a voluntary programme for logging project data, which allows designers, clients and others to compare performance and techniques.

Tenure

Tenure is directly linked to who will be the beneficiary of the retrofit project and therefore willing to invest in it. A retrofitted building will generate direct savings in energy costs and will also have enhanced value as an asset. In most commercial leases at present, savings in energy costs will accrue to the tenant and so owners, who are realistically the only people able to invest in such projects, may have relatively little interest in retrofit. Richard Francis deals with this important and complex issue in detail in Chapter 3. Designers must fully inform themselves of the tenure, service charge and energy cost responsibilities in their retrofit projects.

Age

The age of a building is not, in itself, a factor that directly influences retrofit measures. It is not necessarily even a guide to condition, as there are perfectly sound 200-year-old buildings and unsound 20-year-old buildings. The main diagnostic use of age will be in making an initial assessment of the likely structure, fabric and materials. In older buildings, age will also play a bigger part in historical significance and status, which will restrict the choice of retrofit measures.

Condition

If a building, or an element of the building (e.g. the cladding) is in a particularly poor state, a decision will be made early in a project whether to refurbish or demolish and rebuild. Danger of structural collapse will affect only a tiny fraction of the projects in the UK and even then it is likely that some parts of the structure can be saved if there is an architectural reason and/or an economic or carbon-saving case for doing so. The worse the condition of a building being refurbished for reasons other than energy efficiency (and therefore the larger the amount of alteration and repair), the more viable it may prove to implement a low energy retrofit at the same time.

Historical significance and context

There are over 460,000 listed buildings in the UK. Almost all buildings built before 1840 are listed. There is the further category of buildings that planning authorities consider to be of local importance. There are also over 10,000 conservation areas in the UK. An area of expertise relating to energy efficiency and the conservation of historic buildings is developing, about which English Heritage and other national conservation authorities have produced a growing body of publications.

While it may initially seem that all historic buildings do is present constraints on the choice of retrofit measures, in reality the fabric of many old buildings is inherently able to moderate thermal comfort with minimal energy use. That is simply because they were built at a time when resources and technologies for creating comfort were limited. It is particularly important to understand the performance of this fabric thoroughly and precisely, so that retrofit measures do not result in poor outcomes, such as condensation, rot or frost damage due, for example, to poorly designed and installed insulation.

There are certainly some effective measures that are less likely to be usable in historic or conservation area buildings; external insulation and visible plant for on-site renewable energy generation, for example. Planning considerations may rule out external insulation even for buildings of no merit in themselves, because of a presumption about the context, such as the perceived need to maintain consistency of external appearance with neighbouring buildings. Certain buildings possess a fabric that cannot be altered. In such cases, the retrofit project may be limited to optimising the mechanical and electrical systems and possibly generating renewable source energy. Three of the case studies in this book concern listed buildings of various periods.

RETROFIT AND BUILDING INFORMATION MODELLING

Building Information Modelling (BIM) brings together a powerful set of tools that promises much for the design, construction and operation of buildings. A crucial aspect of BIM is its inbuilt tendency to promote integrated multi-disciplinary working from the start of a project – an important condition for low energy design. If effectively deployed, BIM will play a significant role in retrofitting the existing building stock, given the huge scale of the task.

The workflow for a BIM retrofit project promises an exciting and rewarding process, in theory. The design team starts by modelling the existing building, not only in its three physical dimensions but also incorporating energy and environmental performance together with local climate data, site conditions and other attributes. The team can then model any number of design options and explore their energy, costs and other impacts, eventually transmuting a model based on survey information into a model that will produce the construction information. However, there are a few problems to be solved; for example, how to deal with the inherent irregularity of existing buildings. Even if the drawings from which they were built were entirely regular, setting aside consideration of older buildings with numerous accretions, existing construction has a wide range of tolerances. A building element-based, or object-based, BIM model embeds the most data, but if it accurately records every irregularity it will be hugely time and memory consuming. Alternatively, if the walls, for example, are modelled as accurate carved solids rather than as walls with a set of attributes, they will yield less information. With the current state of the art, it is probably best to model existing buildings simply, not try to record every quirk and accept that the BIM will not give you a perfect fit with existing construction. As digital data capacity increases, it will become possible to fully capture the dimensional information of existing construction.

Retrofit projects require far more extensive surveys than building anew. Once upon a time, architects positively welcomed the surveying task as they could assimilate a great deal of other information about the building's qualities while surveying. However, for building above small scale, the job is now generally given to a specialist topographical survey company. A manual survey, even with laser measures and levels, is expensive and time consuming and digital alternatives are coming into the market. The most promising technologies currently are photogrammetry and point cloud surveying.

Photogrammetry uses digital photographs together with a few reference dimensions to rapidly produce a 3D model of a building. Currently, these models cannot be compared precisely with conventional topographical surveys, but they can be adequate for representation and energy modelling with workable accuracy. Point cloud surveying uses a laser device mounted on a tripod adjacent to or inside a building to create a field of points in space which can be imported into a building information model to create vector information about the building and its spaces. As generating and holding building information models of existing buildings becomes an essential part of operating and maintaining built assets, these and possibly other rapid surveying techniques will become the solution to tackling the sheer quantity of data to be captured.

THE ESSAYS

The essays that follow start with Bill Bordass' cautionary tale, 'Energy Performance in Use and Government Policy'. So much of this subject, designing buildings to conserve energy, is simple and the complex bits are fairly well understood. But there are gaps in the thinking, as Bordass has been pointing out longer than anyone. The 'two steps forward, one step back' nature of public policy is perhaps just something the industry has to accept. The main lesson is to rely less on Government; there is much to be done otherwise.

It is widely accepted that the available finance is a greater barrier to a large scale retrofit programme than the availability of technology. In 'Spend to Make: Financing Commercial Retrofits', Richard Francis identifies the trends that are leading a number of companies to invest in low energy retrofit. He analyses models of retrofit project financing, including likely returns on investment, payback periods, the difficulties of factoring for the long term,

such as uncertainty about energy prices, and the role of governmental stimulus measures. He concludes that environmental and financial performance are becoming inseparable, a scenario that offers the best prospect for a large scale programme of commercial retrofit.

Roderic Bunn's essay, 'From Post-Mortem to Life Support: Building Performance Evaluation as a Design Tool', is a practical guide to targeting the retrofit project to achieve the intended savings in energy. His extensive experience of systematic evaluation of building performance leads him to caution against over-measuring and, indeed, an overly process-based approach. Instead, common sense, design intelligence, a few penetrating metrics and a focus on performance will go a long way to achieving the desired but elusive high energy savings.

It is common to see and hear people making claims for the post-retrofit performance of projects with little evidence and sometimes even without hard figures. In 'Evaluating Retrofit Performance: A Process Map' Rajat Gupta and Matt Gregg approach the problem from a more academic angle, from the vantage point of having evaluated the performance of a large number of domestic and non-domestic buildings. They present a rigorous process, developed for this book, specifically for evaluating non-domestic buildings undergoing retrofit.

In the final essay, 'Non Domestic Retrofit: Projects in Germany and the USA', Mark Siddall, who was an early adopter of the German Passivhaus Standard for energy-efficient design of buildings, takes a selective look at the contrasting approaches being adopted in the two countries, both of which also differ from those favoured in the UK. The prevalence of the Passivhaus Standard in Germany sets the bar for energy performance much higher than in the UK, with design for retrofitted buildings aiming to achieve energy use comparable to building anew. However, the paucity of in-use data makes this difficult to substantiate: it is still a work in progress. In the USA, the commercial sector is leading on energy efficiency retrofit, driven by energy cost savings and not by policy or regulations. The wide availability of benchmark data is accelerating this process because the pressure for disclosure means that no commercial landlord or developer wants to be seen as a laggard.

RETROFIT KEYS TO SUCCESS

1 Develop a brief that integrates energy efficiency measures with improvement in architectural design quality. In historic buildings this might be more a process of removing earlier dissonant additions and clutter than a remodelling exercise. Identify the tangible benefits of integration, such as increased comfort, higher performance, a nicer and more functional space.

2 Get an accurate picture of pre-retrofit energy use to establish the baseline. This need not require extensive sub-meter data or fine detail. Start with accurate (actual) billing data followed, where possible, by an assessment of electricity consumption by end use: heating, fan power, pumping and lighting, and all identifiable plug loads (ICT, servers, local fan heaters, etc.).

3 Design for usability, manageability and maintainability. Involve and enthuse the users where possible, and as early as possible. People will make the ultimate difference in energy use. Good design will clarify the design intent and help to minimise the effort required to make the building run well.

4 Model a range of fabric options and mechanical and electrical (M&E) system options, from easy wins to marginal measures and deep retrofit, in terms of energy performance and costs. Calibrate the model with actual (pre-retrofit) energy use. Make the numbers simple enough for everyone to understand and to compare with other buildings and benchmarks.

5 Use life-cycle costing and increase in asset value, as well as energy savings, in assessments of viability/feasibility.

6 Adopt the Soft Landings process from project inception to post-completion aftercare. Put occupant satisfaction and building management centre stage in the design of the systems. The project does not end with the completion of the retrofit contract – it starts running. The outcomes will depend heavily on how the building is used, maintained and managed.

7 Decide which measures to implement as part of the refurbishment project and make a plan for possible future implementation of further measures. Be careful about prioritisation. Often, a lot can be done to improve energy performance with few changes to the fabric. However, if the fabric is to be upgraded, it may be best to undertake a more complete and permanent set of improvements to provide long-term benefits, and to make strategic provision only for things that can more readily be retrofitted (e.g. photovoltaic panels).

8 Ensure that the building and its systems are constructed as designed, and constructed well. Towards completion, bear in mind that proper commissioning of the systems, and induction of users and managers, is critical to performance.

9 Monitor in-use performance by checking energy bills and meter readings, and identify any areas of higher energy consumption and the reasons for it, such as greater intensity of use, longer hours of operation, poor controls, whether automatic or manual or wasteful running. Conduct good post-occupancy evaluations in accordance with the Soft Landings process to gain a full picture of the building's post-retrofit performance, and the need for any interventions to improve performance.

10 Exchange knowledge. Capture and publish the feedback so that future clients and projects will benefit from your insights. Keep up to date with, but also closely question, emerging knowledge, such as the use of BIM for the retrofit design, construction, operation and management of buildings, and design methodologies such as Passivhaus.

ENDNOTES

1   The UK's Sale of Goods Act 1979 states that goods for sale must be fit for purpose, which is a powerful concept to ensure quality standards of appliances with a limited and clearly defined purpose.

2   *Climate Change 2013: The Physical Science Basis*, Intergovernmental Panel on Climate Change. Available at www.climatechange2013.org (accessed 26 November 2013).

3   *Low Carbon Entrepreneurs: Engines of Growth*, Carbon Trust 2013. Available at www.carbontrust.com/media/310425/low-carbon-entrepreneurs.pdf (accessed 20 November 2013).

4   The United Nations Framework Convention on Climate Change (UNFCCC) gives a global greenhouse gas emissions budget of 2 tonnes per person per year (CO_2e) to avoid catastrophic climate change (namely, a mean global temperature rise of no more than 2°C by 2100). The UK's 80% reduction figure, which is similar to that for most of Europe) comes from the 1990 average annual GHG emissions of approximately 10t. For comparison, some other countries' per capita emissions are: Kenya 0.5t, India 1.5t, China 5t, the USA and Australia 20t and the UAE 40t.

5   $MtCO_2e$ stands for million metric tonnes carbon dioxide equivalent – taking into account the greenhouse impact of methane and other gases as well as carbon dioxide.

6   The UK Green Construction Board has adopted the nomenclature of 'Operational' and 'Capital' carbon (Opco and Capco, respectively).

7   Energy from fossil fuels consumed in the construction and operation of buildings accounts for approximately half of the UK's emissions of carbon dioxide. 'Section 2: Construction and Sustainable Development', *Plain English Guide*, Constructing Excellence (2008), p. 5. Additionally, emissions from transport within cities are heavily influenced by urban form and infrastructure.

8   *Low Carbon Construction Innovation and Growth Team Final Report*, HM Government (2010).

9   B. Gething and K. Pucket, *Design for Climate Change*, RIBA Publishing (2013) – a very useful manual for both new-build and retrofit projects.

10 K. Korytarova and D. Urge-Vorsatz, *Risks and Opportunities of Building Retrofit: Retrofitting Public Buildings in Hungary*. Available at www.salford.ac.uk/__data/assets/pdf_file/0005/142475/099-Korytarova.pdf (accessed 20 November 2013).

11 *Energy Benchmarks TM46*, CIBSE (2008).

chapter 2

BILL BORDASS

ENERGY PERFORMANCE IN USE
AND GOVERNMENT POLICY

INTRODUCTION

It is the use of buildings, and not the buildings themselves, that expends energy and causes greenhouse gas emissions. Yet Government policy and industry practices have consistently focused on the design and construction of buildings, not what happens once they are handed over. The story of the highly promising idea of Display Energy Certificates shows how a real opportunity to make significant cultural change was lost. The salutary lesson is that the design and construction industry must take the initiative to change practices, starting with a true understanding of how its products actually perform in operation.

BACKGROUND

The potential for reducing energy demand in buildings is widely recognised. However, while low energy use is claimed for many new buildings, their actual performance often falls well short of design estimates. Annual electricity use can easily be twice the predicted level, while energy consumption for heating varies widely. While the general expectation of the public and the policy-makers is that new must be better, some new and refurbished buildings have higher carbon emissions than their much older predecessors. Fortunately, as the evidence builds (for example, see www.carbonbuzz.org), there is a growing realisation that these 'performance gaps' really do exist. However, the construction industry and Government are somewhat befuddled as to why and what to do, with committees now pondering matters and threatening to make things complicated and bureaucratic.

Case study evidence of performance gaps has been around for many years, including some publications by the author over ten years ago[1, 2] (see Figure 2.1). Sadly, those with the power to change things tended to ignore the warning signals or to dismiss them as anecdotal, at least until very recently. An important reason for this blind spot is that, over the years, neither Government, nor the design professions, nor the construction industry has invested nearly enough in understanding building performance in use and developing it as a knowledge domain, as has been argued by Frank Duffy, past President of the Royal Institute of British Architects (RIBA).[3, 4]

Policymakers tend to make the category error that building performance is largely about construction and regulation, not the result of a much wider range of influences, as buildings come into being, are occupied and evolve through time. They also tend to look to the construction industry for solutions. The thinking is reflected in the titles of Government reports and initiatives, including the Egan Report *Rethinking Construction* (1998),[5] the Fairclough Report *Rethinking Construction Innovation and Research* (2002)[6] and in the naming of the Green Construction Board (2011).[7]

2.1 The cover illustration by Louis Hellman for the author's 2001 publication *Flying Blind*. This shows the designer, builder, facilities manager and owner of a recently completed building all ignoring the evidence of a big difference between estimated and actual performance, what is now known as the performance gap. (The data for the graph shown came from a building that had won a sustainability award.) The publication advocated using energy certificates to disclose actual performance and motivate action. It also expressed concern about the consequences of the fragmentation of the buildings and energy policy that had previously been concentrated in the Department of the Environment

In 1970, the UK Government established a Department of the Environment (DoE), which included the former ministries of Housing and Local Government, and Public Building and Works, so bringing together many of Government's building-related activities. In 1992, DoE also took over the Energy Efficiency Best Practice programme. For a brief period, DoE was a focal point for buildings and energy research. In 1997, things began to disintegrate, starting with the ill-considered privatisation of the Building Research Establishment (BRE), which reported to the DoE. Following the 2001 election, DoE's successor, the Department for Environment, Transport and the Regions (DETR), was further dismembered, eventually ending up as the Office of the Deputy Prime Minister (ODPM). Amongst other things, wider environmental matters went to Defra (the Department for Environment, Food and Rural Affairs), the Energy Efficiency Best Practice programme to the Carbon Trust, while DoE's responsibility for construction sponsorship shifted to the Department of Trade and Industry (DTI, today called Business, Innovation and Skills, BIS). In 2008, the Department of Energy and Climate Change (DECC) was also established. Fragmentation between so many departments has led to confused and disjointed policies about energy and buildings, with no common technical core.

In 2002, the Fairclough Report[8] considered the implications for construction research of the completion of BRE privatisation, with the ending of Government's five-year transitional arrangement, and the transfer of construction sponsorship from the Department of Transport, Local Government and the Regions (DTLR – the successor to DETR) to DTI. The report regarded innovation and research as largely to do with construction and, consequently, the responsibility of the construction industry, which would have to vie with other industries for Government support. That soon led to the closure of the Government's specifically buildings-related research programme, Partners in Innovation.

Meanwhile, the Fairclough Report saw building performance largely as a matter for regulation, with little wider implication or reach. It did, however, identify four roles in which it would be in the Government's interest to fund building research directly: those of regulator, sponsor, client and policymaker. This research would relate to 'issues that go wider than the construction industry': specific mention was made of climate change, energy efficiency and unforeseen circumstances.

Sadly, and in spite of all the evidence, it has been difficult for policymakers to appreciate that building performance concerns much more than construction, and to achieve joined-up Government thinking and action. A recent shaft of light has been the Technology Strategy Board's sponsorship of a programme of about 100 building performance evaluations, which are referred to in the essays by Roderic Bunn, and Rajat Gupta and Matt Gregg. This programme has a finite life, ending in 2014. To avoid a glut of unintended consequences, there needs to be a continuing flow of performance feedback information in the public interest, providing data, connections and insights to support the radical improvements to policy and practice that will be required.

BUILDING PROFESSIONALS AND BUILDING PERFORMANCE

Where does this leave the building professional? To protect society's wider interests, and in return for their protected status, professionals are 'granted the privilege to think' (to use the words of a former chair of the Construction Industry Council, Keith Clarke) and have a responsibility to 'do the right thing' (to quote from the Charter of the Institution of Civil Engineers), going beyond their obligation to whoever pays their fee. The challenges of sustainability now bring professional obligations into sharp focus, with the common interest now at the global scale too. As Malcolm Bull puts it: 'climate change does not tempt us to be less moral than we might otherwise be; it invites us to be more moral than we could ever have imagined'.[9]

A milestone in the history of building performance was the book of the same name,[10] published in 1972 by the Building Performance Research Unit at the University of Strathclyde. History has shown this to have been more epitaph than manifesto. In the same year, Stage M (Feedback) was removed from RIBA's document *Architect's Appointment*, on the grounds that the service could not readily be quantified and clients were unwilling to pay for it. Sadly, this included government clients. However, at the time, government departments still had their building professionals, works departments, research units and the Building Research Establishment, and so had been doing a lot to close the feedback loop, implicitly and explicitly. In the ensuing decades, Government tended to outsource, privatise or abandon these activities, but neither industry nor the building professions put effective alternative feedback systems in place.

Without such feedback, how can building professionals know that they are doing the right thing? Frank Duffy has said: 'Plentiful data about design performance are out there, in the field … Our shame is that we do not make anything like enough use of it'.[11] Because such follow-through and feedback is far from routine, even now, many people say it can't be afforded. On the contrary, we can't afford to neglect it. Without routinely following through into use and feeding back the experience, how can we test and refine our proposals? We might even end up not improving performance at all, let alone to the radical extent that policymakers have been anticipating.

Professional institutions already require their members to understand and practise sustainable development: surely this must include understanding the outcomes of their own activities? In recent years, things have at last begun to move in this direction: for example, the RIBA Plan of Work 2007 incorporated Stage L (Post practical completion); and two of the RIBA Plan of Work 2013's seven stages relate to use: Stage 6 (Handover and Close out); and Stage 7 (In Use). However, the necessary follow-through and feedback activities are not yet well defined or widely practised.

FIVE STEPS TOWARDS BETTER PERFORMING BUILDINGS
Keep things simple and do them well
Studies in the 1990s, including the PROBE (Post-occupancy Review Of Buildings and their Engineering) series of published post-occupancy evaluations (POEs),[12] revealed that unmanageable complication was the enemy of good performance. At the same time, many basic things one would hope to be able to take for granted (e.g. the thermal integrity of the fabric and the functionality of manual and automated controls) often left much to be desired. The buildings that worked really well tended to have received careful attention to detail: in design, during construction, and before and after handover. Another important ingredient of good performance was an individual (or, better still, several individuals) committed to getting a good result: process alone was no substitute for this leadership.

Robust, not fragile buildings
With dedicated input, complicated buildings can also work well if sufficient effort is put into both their procurement and their management; from briefing through design and construction and on into operation. As PROBE and other POEs have found, the best-performing buildings of this type often had a dedicated client representative who had provided the necessary leadership and insight right through the process. However, as time passed, the performance of some complex buildings that had worked well when monitored in their early lives deteriorated badly; for example, when economic changes caused maintenance and management budgets to be cut, or if skills and understanding were lost when facilities management was outsourced. Better to be simpler and more robust, particularly in the case of public buildings, as more complex tends to mean more fragile. Sadly, over the past decade, buildings and the related legislative requirements have headed off in the opposite direction, becoming ever more complicated. Examples include recently constructed schools: expensive to build, expensive to occupy, and often with large performance gaps not just in terms of energy and carbon, but for occupant satisfaction as well. Theory tends to favour the more complicated

solution over the simple one, but performance in use points to the importance of making things robust, usable and manageable, and paying close attention to detail.

Improve the process

The concept of completing work, handing it over and going away immediately is not fit for purpose for today's buildings. Indeed, the whole procurement process needs to be re-examined, sharpening the focus on clear outcomes from inception right through into use. At present, unfortunately, rather than being maintained and nurtured, the golden thread from design intent to reality is frequently severed as a project moves from stage to stage, sometimes with an almost complete change in players. Given such discontinuities, it is inevitable that performance gaps will open up, targets will be missed, innovations will not work as anticipated, and lessons will not be learned from unintended outcomes.

To help bind things together, the Soft Landings Framework[13] has been developed to allow any project, in any country, with any procurement system, to give more emphasis to outcomes. It reinforces existing processes at five critical stages:

1. inception and briefing
2. managing expectations during design and construction
3. preparation for handover
4. initial aftercare, and
5. longer term aftercare, typically for three years after handover.

The approach works best if one or more members of the project team adopt the role of Soft Landings champions, to help to maintain the focus on outcomes and to support and challenge other team members.

Count everything

Designers tend not to have been very good at estimating actual energy performance in use. Indeed, many have preferred to shelter behind the argument that their calculations are to compare options, not to make predictions. The architect has too often asked the computer modeller or building services engineer: Does it meet the regulations? If the answer is yes, the design proceeds; if not, options are reviewed and changes are made – often adding complication, because this tends to make the sums work better, though not necessarily the building itself. The results of the calculations are often difficult to understand. They have also tended to focus on so-called 'regulated loads', representing the energy end-uses covered by the Building Regulations, i.e. heating, hot water, cooling, ventilation and fixed interior lighting. Moreover, the estimates tend to assume standardised conditions. The numbers for energy use may look good, but the assumptions can be questionable. Often the forecasted consumption is just the tip of the iceberg, particularly in non-domestic buildings, where the energy used by the occupier's equipment and management can easily predominate. Unfortunately, many building designers regard this as nothing to do with them. In practice, however, if the priorities are communicated clearly and early, and the likely outcomes are monitored and managed throughout the procurement process, dialogue can be highly influential. It allows occupiers to take more seriously the specification of their own equipment (e.g. computer and catering equipment); how they use and manage their building; and any support services they engage – all of which can have major effects on in-use performance. Continuing reviews and conversations as a project proceeds will also help designers to make their building and systems more capable of being controlled and managed effectively in relation to the likely patterns of use.

Focus on performance in use

In 2001, in the publication *Flying Blind*,[14] the author argued that building performance needed to be made visible to spur people into action. If the owners and occupiers of a building were required simply to disclose the annual energy used in operation, this would provide a non-punitive way of starting the transformation to better building energy performance in use. An opportunity came in 2003, when the EU's Energy Performance of Buildings Directive[15] led to the development of Display Energy Certificates (DECs) for non-domestic buildings, based on actual energy use. DECs came into force in England and Wales in October 2008, starting with public sector buildings of over 1,000 m², but have recently been extended in a half-hearted and confusing manner;[16] for example, by requiring eligible commercial buildings to display their theoretical and not achieved performance. Sadly, while DECs have helped to expose the energy performance gap, they have not achieved anything like their potential as a cornerstone for energy and carbon performance improvement. An important reason, discussed below, is that policymakers have not invested in the infrastructure to support DECs properly, or to integrate them with other buildings and energy policy measures, of which there are now far too many.

HOW NOT TO PURSUE POLICY
– DISPLAY ENERGY CERTIFICATES (DECs)

After the PROBE project had published its first 16 reviews of the performance in use of recently completed buildings, the team obtained Government funding to review the results[17] and consider the next steps. One outcome was the decision to apply for EU research funding to extend the approach to Europe, with partners from Belgium, Denmark, Germany, Greece, Sweden and the Netherlands. In 2000, a bid for EuroPROSPER (EU PRoject for Occupant Satisfaction, Productivity and Environmental Rating) was rejected as being too ambitious. A successful resubmission was made by the project leader ESD in 2001, with the scope reduced to offices and concentrating on an operational energy rating and a much simplified assessment of occupant satisfaction.

While the resubmission was being prepared, the Energy Performance of Buildings Directive (EPBD) was progressing through the European Parliament, including proposals for building energy certificates. The revised EuroPROSPER

submission argued that the project could pave the way for building energy certificates based on actual energy use, which would in turn lead to wider interest in other aspects of in-use performance. The research was carried out in 2002–04, with Defra providing UK matching funds through the vehicle of the newly established Carbon Trust. The power and usability of the demonstration energy certification software developed for offices surprised even its originators: it could not only benchmark energy performance automatically but, from a small amount of data, provide an estimated breakdown into end uses, together with an indication of likely improvement measures, including typical costs and savings. These initial estimates could then be fine-tuned by the assessor as necessary, with the software taking care of the calculations.

When it was finally ratified at the end of 2002,[18] the EPBD put more stress on calculated energy ratings. The mandatory requirement to display a certificate was also restricted to public authority buildings and buildings frequently visited by the public of over 1,000 m² in usable floor area.

The EPBD's introductory Recitals stressed the great unrealised potential for energy savings, the importance of managing energy demand, the need for regular certification for public buildings, and for certificates to describe 'the actual energy performance situation to the extent possible'. However, apart from the requirement for regular inspections of boilers and air-conditioning installations, the main Articles focused on investment measures and theoretical calculations and contained relatively little on operational measures and actual energy use.

The EuroPROSPER team nevertheless argued successfully in the UK and Europe that, while Energy Performance Certificates for new and empty buildings could only be based on modelled energy performance, for public and commercial buildings in operation, certificates based on actual energy use would be more revealing and cheaper to produce. They could also help to save energy quickly and cheaply, by motivating management to make year-on-year improvements. The energy data required could potentially be updated automatically by the utility companies, and the results aggregated into portfolio statistics at an organisational level. The EC asked CEN, the European Committee for Standardization, to develop supporting standards for the EPBD: its outputs included standards for both Asset (calculated) and Operational (measured) Ratings.

In the UK, the case for energy certificates based on Asset and/or Operational Ratings was recognised in ODPM's 2004 consultation document for England and Wales,[19] where the EuroPROSPER proposals were widely referenced, including the need for better benchmarks and effective integration with utility metering and billing. One disappointment for the advocates of energy use disclosure was that, while the EuroPROSPER team had proposed a single certificate that showed both Asset and/or Operational Ratings in a transparent manner (see Figure 2.2), and CEN had endorsed it as an option in its draft standard, which became BS EN 15217:2007, ODPM's consultants advised that they should be separate items. Apart from that, the prospects for developing a good Display Energy Certificate (DEC) system looked encouraging, and the approach was also endorsed in the consultation responses. However, the scheme soon ran into difficulties owing to the fragmentation of policymaking about buildings and energy, as outlined below.

The EuroPROSPER team had proposed a substantial investment in benchmarking, to extend the 'tailored' system used for offices to the other public buildings that were the initial focus of display requirements in the EPBD, especially education, health and sports. If the UK had pioneered it, the system could potentially have been adopted across the EU, and perhaps beyond. This might also have had economic benefits for the country but, disappointingly, the international dimension was of no interest to those departments and agencies that had no remit outside the UK.

ODPM said that it could not invest in developing a system and the associated benchmarks until its consultation was complete and a decision had been made on whether or not to proceed with DECs.

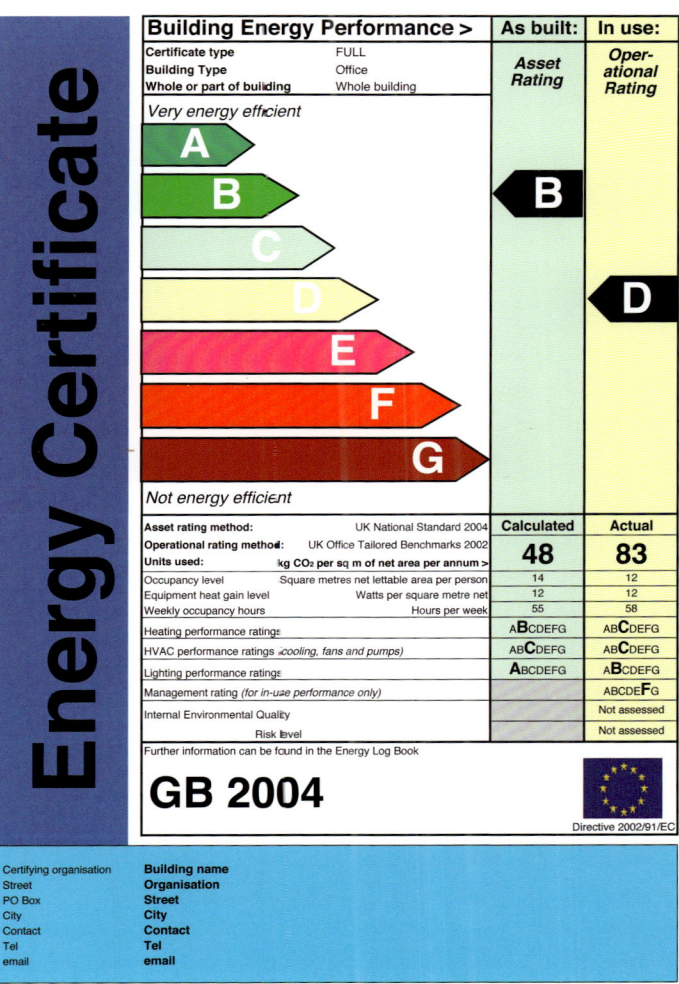

2.2 Proposal by the EuroPROSPER team in 2003 for a two-column energy certificate graphic showing both Asset (calculated) and Operational (measured) energy ratings. A second page gave more technical detail

Although the Carbon Trust had taken over the benchmarking publications from the Energy Efficiency Best Practice programme, at the time it was not interested in benchmarking, also arguing that its remit was not to overlap with what the Government was doing, that certification was ODPM's responsibility and it was not the Carbon Trust's job to prepare the ground for it.

Other funders or supporters were not prepared to put money into benchmarking, unless they could be given some certainty about whether and how the Government was going to use it.

The gas and electricity regulator, Ofgem, told ODPM that to get gas and electricity billing into good shape to feed into DECs would be an unfair burden on the utilities. Instead, they saw it as a service that individual customers should request and pay for.

With tailored benchmarking proving impossible to fund in 2004, the Usable Buildings Trust then proposed an approach that demonstrated how to get started on DECs with rudimentary benchmark data.[20] This approach was then used to revise the second EU research project, EPLabel. ODPM offered financial support, starting in April 2005. Unfortunately, an election was called, and the decision had to be deferred pending a new Government. Although the same party stayed in power, the incoming Minister was sceptical about the idea of DECs on the grounds that two types of Energy Certificate was 'gold plating' an EU Directive.

Fortunately, in June 2006, ODPM (now called CLG, Communities and Local Government, in yet another UK Government department change of name and function) decided that it did make sense to have DECs and to support EPLabel. However, with 14 countries involved and only a few months left, a massive opportunity was lost for CLG to shape the system. It was also not prepared to adopt and adapt the system that had been developed, seeing that as anti-competitive. Instead, EPLabel helped them with performance requirements and public consultations.

Owing to all the delays, the implementation of an energy disclosure system in time to meet EU deadlines had now become urgent, with a Display Energy Certificate system to be introduced in early 2008, becoming mandatory from October. This meant that benchmarks were needed rapidly, a task assigned to the Chartered Institution of Building Services Engineers (CIBSE), but with no government budget. With the agreement of key stakeholders, simple placeholder benchmarks were developed and published in CIBSE TM46 *Energy Benchmarks*, with the expectation that, once the DEC system was in operation, the Government would provide funds to develop the benchmarks. Unfortunately, at the time of writing, five years later, no funds have been forthcoming, so the whole enterprise of improving building energy and carbon performance rests on insecure foundations.

While DECs have helped to expose the performance gaps, their implementation has been a disappointment, for three main reasons:

1. The Government seems to regard them as a drag on economic growth, not an evolving window on real energy performance and the anchor for a whole variety of policy and other measures.
2. Despite their importance in providing clarity of communication and furthering of policy objectives, there has been no Government investment in benchmarking for a decade.
3. DECs have not been extended to private sector buildings, in spite of strong support from influential bodies, including the Confederation of British Industry. Partly, this is because of concern about the benchmarks.

CONCLUSION

For all the policy interest in improving building energy and carbon performance, we still lack clarity about the key objective: How is this building actually performing? We also lack a set of joined-up policy instruments that can concentrate the actions of all the players involved, from investors through to maintenance contractors, on purposeful improvement and help them to work together.

The situation has been exacerbated in the UK, because policymakers regard going beyond the letter of an EU Directive as 'gold plating' and to be avoided. This line of thinking was powerful for the previous Government, but is pathological under the present one, which has a policy to 'copy through' the clauses of a Directive into British law, without enhancement. This myopic approach creates a confusing jumble of bureaucratic requirements, instead of a well-integrated set of policy measures that can adapt themselves to accommodate new Directives. Instead of converging onto understanding and improving building energy use in operation, our policies circle hopelessly around it.

Whatever Government does, building designers need to become much more familiar with how their buildings work in use. Only then will they understand what they really need to do to improve performance outcomes. Some assistance is now available from Carbon Buzz (www.carbonbuzz.org) which has been developed with support from the Technology Strategy Board, RIBA and CIBSE. This platform allows people to deposit and share their design and in-use energy data and to identify contributors to the performance gaps. However, to make real progress, we need much more consistent and effective integration between industry and policy measures for reporting and benchmarking building energy performance.

ENDNOTES

1   *Building Research & Information, Special Issue on post-occupancy evaluation*, Vol. 29, Issue 2 (March–April 2001). www.tandfonline.com/toc/rbri20/29/2

2   B. Bordass, *Flying Blind – Everything You Wanted To Know About Energy Use In Commercial Buildings But Were Afraid To Ask*, Association for the Conservation of Energy (October 2001).

3   F. Duffy, 'Linking Theory Back To Practice', *Building Research & Information*, Vol. 36, Issue 6, 2008, pp. 655–758.

4   F. Duffy, 'Reflections on Stage M: The Dog That Didn't Bark', in S. Mallory-Hill, W. Preiser and C. Watson (eds), *Improving Building Performance*, Chapter 26, pp. 315–320, Wiley (2012).

5   Egan Report *Rethinking Construction* (1998) www.constructingexcellence.org.uk/ pdf/rethinking%20construction/rethinking_ construction_report.pdf (accessed 27 January 2014).

6   Fairclough Report *Rethinking Construction Innovation and Research* (2002) www.berr.gov.uk/files/file14364.pdf (accessed 27 January 2014).

7   Green Construction Board (2011). The Green Construction Board was set up in 2011 by the Department of Business Innovation & Skills to promote carbon reduction in and by the property infrastructure and construction industry.

8   J. Fairclough, *Rethinking Construction Innovation and Research*, Department for Transport, Local Government and the Regions, DTLR (2002).

9   M. Bull, 'What Is The Rational Response?', *London Review of Books*, Vol. 34, No. 10, pp. 3–6 (24 May 2012).

10 T. Markus, P. Whyman, J. Morgan, D. Whitton and T. Maver, *Building Performance*, Applied Science Publishers, London (1972).

11 F. Duffy, 'Linking Theory Back To Practice', *Building Research & Information*, Vol. 36, Issue 6 (2008), pp. 655–758.

12 *Building Research & Information, Special Issue on post-occupancy evaluation*, Vol. 29, Issue 2 (March–April 2001).

13 M. Way, W. Bordass, A. Leaman and R. Bunn, *The Soft Landings Framework*, BSRIA BG 4/2009, BSRIA and the Usable Buildings Trust (July 2009).

14 B. Bordass, *Flying Blind – Everything You Wanted To Know About Energy Use In Commercial Buildings But Were Afraid To Ask*, Association for the Conservation of Energy (October 2001).

15 Official Journal of the European Communities, *DIRECTIVE 2002/91/EC OF THE EUROPEAN PARLIAMENT AND OF THE COUNCIL of 16 December 2002 On The Energy Performance Of Buildings*, L1.65-71 (4 January 2003).

16 A. Warren, *DCLG Takes Top Certificate in Creating Confusion, Energy in Buildings and Industry*, Vol. 12 (February 2013).

17 *Building Research & Information, Special Issue on post-occupancy evaluation*, Vol. 29, Issue 2 (March–April 2001).

18 *Ibid.*

19 Office of the Deputy Prime Minister, *Proposals for Amending Part L of the Building Regulations and Implementing the Energy Performance of Buildings Directive* (July 2004).

20 Usable Buildings Trust, *Onto The Radar: How Energy Performance Certification and Benchmarking Might Work for Non-domestic Buildings In Operation, Using Actual Energy Consumption* (June 2005).

Level 11
Level 10
Level 09
Level 08

Level 07
Level 06
Level 05
Level 04

Level 03
Level 02
Level 01

chapter 3

RICHARD FRANCIS

SPEND TO MAKE
FINANCING COMMERCIAL RETROFITS

INTRODUCTION

Both the how and the why of financing energy retrofits are under review. Traditional barriers to inaction, mainly financial and attitudinal, are eroding as the value (not cost) of making properties energy efficient becomes paramount. There will be continuing resistance to financing retrofits until it is clearly demonstrated that they not only reduce energy but, in doing so, increase asset worth. But the new threat of obsolescence caused by poor energy performance is already allaying long-held suspicions and loosening purse-strings.

FACING PAGE 199 Bishopsgate, London. See Case Study 1

23

INTRODUCTION

It is widely accepted that lack of available finance is a greater barrier to large-scale retrofit programmes than the availability of technology. This belief, while broadly true, needs to be refined. There is money available for energy retrofits (and it is increasing all the time), but there remains a reluctance to act. Money has been the primary barrier, not so much because there was too little to spend but because there was too little to earn.

Despite this commercial reluctance, it is clear that something must be done about the existing stock of poorly performing buildings. Whatever may be said about strict carbon regulation for new buildings, one thing is certain – it is not going to solve the carbon problem. With replacement rates at 1–1.5% per year, about 70% of existing buildings will still be in use in 2050. About 40% of these buildings will have been built prior to 1985 and the introduction of energy standards in Building Regulations.[1]

This means that there are enormous challenges and opportunities in the UK market. In many respects, the future of UK commercial property is primarily determined by the past, and what can be done to change sub-par energy performance into future viability and adequate long-term performance. With an older building stock, the UK feels the carbon challenge more acutely than other countries, but it also means that large-scale retrofitting is inevitable.

Significant change is under way that is altering the business equation for retrofits. Regulation is forcing transparency and poorly performing buildings are at risk of losing value. At the top end of the market, occupiers have begun to seek out and reward properties that are energy efficient, making owners more likely to consider taking action. In the end, it will not be potential energy savings but the threat of lost revenue that drives retrofits forward. The costs of inaction are slowly beginning to outweigh the costs of action and the business case for retrofits is steadily unfurling.

BACKGROUND

Gaining support for energy-efficient retrofits has been a difficult task. Even when both money and proven technology are available, and future savings are all but assured, organisations still tend to balk. Their reasons are predictable and practical: retrofitting is disruptive to business, often someone else (the tenant) benefits from the investment, predicted energy savings rarely materialise, the asset may be sold before costs are recouped and capital and operating expenses are handled separately within organisations, to name just a few.

In an industry not known for systematic evaluation, there is very little evidence of what actually works. While the industry has invested significant time and resources designing new low carbon properties, it has largely ignored the existing building stock. Professionals still talk in unconvincing terms of 'projected paybacks' largely because they do not have the data to speak with certainty. Owners or investors seeking examples of successful projects to provide a template for a successful energy-efficient retrofit are likely to be disappointed. The evidence that expensive technologies will actually deliver results is, in many cases, simply not there.

When focusing purely on costs and paybacks (rather than added value) proponents of retrofits are playing a losing game. The economics for action often do not add up when based on energy savings alone. It is only recently that energy efficiency has been connected with values beyond lower operational costs. There are many reasons for this shift, including regulation and the transparency that it is encouraging, but the fundamental fact is that energy efficiency is now less about saving the planet (or operational costs) and more about saving assets from depreciation. Within this scenario, actors who were previously reluctant to find money suddenly start to look a little harder. This is precisely the situation we find ourselves in today.

Much has been written about how energy efficiency competes with other objectives for an ever more limited amount of capital. Likewise, observers conclude that retrofits (or any improvements for that matter) are unlikely to be undertaken in the short term as organisations cling to money due to the recent, and ongoing, economic woes. While this is no doubt true for many organisations, it obscures the fact that the energy efficiency issue was on the back burner during the economic good times and that it has never been a priority, even when money was available. What is also noteworthy is that interest and action in retrofits is greater now than it was a decade ago, even though most organisations have less money to spend overall. Something other than money, or the lack of it, is clearly at work.

In an influential industry study, the Better Buildings Partnership (BBP) concluded that it is not a lack of funding that stymies retrofit efforts.[2] Instead, it is the lack of a convincing business case that prevents action. Organisations can make capital available provided there is a compelling reason – but real, convincing economic arguments for retrofits have remained an industry weak spot. What is needed is an approach that is standardised, recognised and predictable so that owners and funders can base financial decisions on strategies that have been shown empirically to be profitable. This strategy requires a combination of finding both technologies that work and financial arrangements that reduce risks and reward the investor.

Funding low-cost improvements with low risk and relatively short payback periods of 3–5 years is not the real challenge. Such measures already satisfy the business case and, therefore, many organisations have already undertaken these improvements. The real challenge for the industry is how to create mechanisms for capital-intensive energy efficiency retrofit improvements.

BARRIERS TO RETROFITTING

In order to satisfy the business case, there must be clear proof that energy-efficient buildings perform better not only in environmental terms but also financially. The 'performance gap' between promised and delivered energy efficiency is well-documented and post-occupancy evaluations routinely show that many 'energy-efficient' buildings fail to deliver. This is a significant issue for the industry and one that must be addressed before investors, understandably, feel comfortable about releasing capital. There are additional questions concerning how energy efficiency is valued in the market. While occupiers appear to be moving towards rewarding energy-efficient properties, there is still a long way to go in providing the kind of proof that owners and investors require in order to act.

Financial

The commercial real estate market is particularly risk averse. When outcomes are uncertain (as they are in energy efficiency projects), owners tend not to respond. Owners are willing to invest in many building features if they have a perceived or demonstrated return on investment. The lack of hard and replicable numbers for energy-efficient interventions has meant that owners have questioned:

1. whether expected environmental returns will materialise, and
2. even if they do, whether they will translate into direct financial returns.

Historically it has been difficult to assess the value of energy efficiency, in part because it was difficult to know how buildings compared and also because no one was particularly interested in finding out. Traditionally occupiers have not shown a preference for energy-efficient buildings, nor was it possible for valuation professionals to incorporate improvements into the overall value of the asset. There was little empirical evidence to suggest that the market was prepared to reward investments in energy efficiency. This stance is changing, but in many markets, and for many investments, energy efficiency is still not a top priority.

It is not just about what works – it is about who benefits. Even if retrofits demonstrated a definitive payback over the long term, it is still unlikely that many projects would be undertaken. Turnover in the commercial marketplace is particularly quick, and short-term thinking prevails. On average, an owner holds a property for about five years, and so any project with a payback period longer than this is likely to be ignored. Large projects may have payback periods that substantially exceed the projected ownership timeframe and owners fear that costs will not be recouped at the time of sale.

This concern explains why more complicated and innovative financing arrangements, such as energy service companies (ESCOs), have been effective primarily in the MUSH (municipal, university, schools and hospitals) market. Here, unlike in the commercial market, owners hold assets over a long period of time and directly benefit from any cost savings. With a clear line of ownership and a clear benefits structure it is far easier to overcome the traditional barriers. The business case for energy efficiency exists; it just does not satisfy the dynamics of the commercial market.

Complicated ownership and lease arrangements further stall change. Many buildings in the commercial market are owned by multiple parties. Reaching agreement on risky projects among all the interested parties is notoriously difficult. Moreover, commercial properties often have lease structures that discourage investment. Although the owner is responsible for financing works, it is the tenant who reaps the cost savings. This 'split incentive', where owners assume the costs but do not reap the benefits, is perhaps the biggest barrier to retrofit projects.

Technical

For retrofits to be an attractive investment option there must be convincing evidence that technologies and best practice options actually work. At the moment, this is not the case. The industry is poorly equipped to provide evidence that retrofits succeed because so little work has been done on evaluating the actual performance of retrofitted buildings.

There are now numerous initiatives under way to review retrofit performance in diverse sectors, and there are many case studies which point to significant financial benefits, often with payback periods that are much shorter than anticipated.[3] But a systematic analysis of different retrofit options and outcomes for the commercial office sectors is lacking. However powerful individual case studies may be, they are not the standardised compilation of evidence that underwriters rely on to make decisions.

Where evaluation of new low carbon buildings has taken place, the results are hardly encouraging. The difference between predicted and actual performance typically varies considerably, often by 200–300%. Owners and investors are particularly wary of numbers such as these, because they are further evidence of the risk involved in a low carbon industry that is still very much in its infancy.

Attitudinal

It is not only owners who have been lukewarm in their response to energy efficiency. Historically, most tenants have been more interested in location and rent than operating costs. Even though energy efficiency has risen up the agenda, there are questions about how deep the market for energy-efficient properties actually is. Certainly, new, prime property and that occupied by top corporate occupiers must increasingly be low energy, but overall this is still a relatively small proportion of the existing stock. What about companies without carbon reporting or corporate social responsibility (CSR) requirements? Will they seek out energy-efficient properties? What about the vast secondary asset market? There is nothing to suggest that the new interest in energy efficiency is as deep as it is wide.

Third-party financiers find themselves in an immature market that lacks the tools that guide other types of investments. Indeed, much of the promising work in retrofits is focusing on developing standards (such as ASTM E 2797-11, Building Energy Performance Assessment Standard) that will make investment decisions on energy efficiency more rigorous, standardised and predictable.[4] Among lenders, there is a requirement for stringent financial underwriting criteria concerning returns on investment and less appetite for large projects with uncertain returns.

Retrofitting is generally viewed as complicated because of the issues and actors involved. It is seen as costly and inconvenient, and its benefits are not transparent. These factors contribute to the perception of retrofitting as being less attractive and profitable than other investment vehicles.

FINANCIAL MODELS

Like attitudes to retrofits, traditional approaches to funding in the UK have been uninspiring in general. Banks and other financiers have kept their distance and the more innovative/complicated financial initiatives found in other countries have gained little traction here. For example, special purpose vehicles, such as ESCOs, have been very successful throughout Europe and the United States but have not taken hold in the UK.[5]

The investment market for retrofits is changing, but the UK still lags behind other countries in developing new arrangements for reducing risk, institutionalising performance guarantees, underwriting and providing new models of financing. In the USA, for example, auditing and underwriting tools are being developed to make retrofitting a mainstream financial activity with a high degree of standardisation and rigour. Any large-scale retrofitting effort in the UK will need to follow the same course to reduce the uncertainties that presently foster confusion and inaction.

Traditional financing

Broadly speaking, companies in the UK have traditionally had three means of funding retrofits – owner financing, lender financing and the creation of ESCOs or other special purpose vehicles. Of these three mechanisms, the first has been by far the most widely employed, with lender financing and ESCOs underutilised here in comparison to other countries.

Owner financing

The majority of energy efficiency projects conducted in the UK fall under this category. Owner financing from capital or operating budgets represents the most straightforward way of paying for energy efficiency improvements. To date, energy efficiency improvements have been of the light-touch retrofit variety with typical payback periods of between three and five years.[6] The fact that less capital is committed, coupled with the assurance that returns will be realised by the present owner, can make internal financing an attractive option for owners, particularly owner-occupiers.

Owner-occupiers typically take a longer view, are willing to accept longer payback periods and accept higher risks when the direct returns are potentially higher. However, owner-occupied properties represent a small segment of the market, making up less than one-third of all commercial properties.

Owner financing faces a number of hurdles whose size increases with the amount of finance needed. Owner financing requires cash reserves and a commitment to energy efficiency in the face of competing priorities. Other company investments have typically taken precedence over energy improvements and, in recent times, companies have been less willing to spend prospectively for future returns.

Under this type of financing, the owner assumes 100% of the under-performance risk. The afore-mentioned gap between design intent and actual energy performance has deterred many owners from taking on larger, more complicated and less certain energy retrofits.

Lender financing

An alternative, but less frequently used, form of funding is financing through banks or other types of lenders. Funding institutions are increasingly being drawn to retrofit projects from which they formerly would have shied away, particularly as the economics of energy efficiency become clearer. Lenders are generally more comfortable with factors like location, occupancy rates and rents that are easier for conventional models to assess.

Lenders have had a difficult time understanding the business model for retrofits. Revenue from future energy savings is regarded as an unusual source of surety for traditional lending. There is an understandable lack of confidence in the ability of projects to deliver actual energy savings.

For most commercial properties, the operational savings that result from energy retrofit projects are often meagre. For larger, deeper, more complicated (and expensive) energy retrofits, the amount saved in energy from the retrofit process is insufficient to repay the debt. Unless retrofits can bring in revenue beyond that generated by energy savings, lenders have difficulty in justifying the loan.

Valuing energy efficiency improvements has been difficult for lenders because historically appraisers have not reflected these in their valuations. This lack of concrete, identifiable value has deterred lenders from taking on investments which are perceived to be risky and insufficiently backed.

This situation, however, is changing. Lenders are beginning to recognise that energy efficiency loans can help to preserve the value of a building by avoiding obsolescence. Because obsolescence is tied to the value of the asset, it is of increasing importance to lenders.

Energy service companies (ESCOs)
An ESCO is a commercial entity that provides holistic energy solutions. Many ESCOs can also assist with project financing (as an active or passive party). Projects are typically large scale with the contract covering a five- to ten-year period or longer.

Under a typical ESCO arrangement, the building owners pay a fee to an ESCO in return for the promise of reduced energy costs. The savings are generally guaranteed to exceed the fee and are designed to pay back the capital investment of the project over time. One of the benefits of this arrangement is that the ESCO assumes the risk: if the return on investment is poor, it is the ESCO that is responsible for any costs that are not recouped.

In the commercial property market, the length of typical ESCO projects has proved detrimental to ESCO growth. Building turnover discourages owners from taking on retrofit projects with longer timeframes. As a result, ESCOs have made limited inroads into the commercial market, although they had more success in public sector settings.

Emerging models of finance
In addition to the three traditional routes to funding, there are new market and Government initiatives under way. Slowly, new third parties, sensing a fresh opportunity wrought by regulation and a changing market, have begun to package products to mitigate the risks that formerly made parties averse to taking action. The Government has played a major part in these new efforts by providing subsidies for energy-efficient equipment and offering 'pay as you save' loan programmes to commercial real estate owners.

Energy service agreements
New managed energy services agreement (ESA) structures are being developed by third parties. An ESA is an arrangement where energy efficiency is sold as a service. It differs from an ESCO because an ESA assumes ownership and maintenance responsibility for the assets over the lifetime of the project. Under an ESA, building owners are not required to arrange their own financing and do not have responsibility for principal or interest payments. Owners do not bear the risk of the performance guarantee, which is sometimes an issue with ESCO arrangements. This is because energy efficiency service providers are compensated only if energy savings are realised.

In a typical example, an owner would pay a fee to the third party energy service provider based on the building's historical energy costs. The third party would make efficiency improvements and pay the actual utility bill, earning its fee from savings generated by the efficiency improvements.

'On-bill' utility financing

Under on-bill' financing (OBF), a utility or a third-party financier provides capital upfront for energy efficiency measures. The owner then repays the investment through an extra charge on the utility bill. Utility financing usually offers low- or zero-interest loans for technologies designed to have a payback of about three years or less.

OBF was first introduced in the USA in the 1990s and has since been adopted by numerous energy providers. Eligible projects include energy-efficient retrofits, such as the installation of LED lights, solar panels or new boilers. OBF creates a 'bill neutral' financing scenario, matching payment terms of the loan to the expected savings. As an example, a £10,000 retrofit project with expected energy savings of £200 per month would be structured to have 50 monthly payments (£10,000/£200) at £200 per month.

Uptake of this offer by private power companies has not been strong. Owners have been wary of yielding finance collection and distribution to utilities. Many utilities have also been reluctant to serve as a loan originator and collector.

Equipment lease finance

Under this arrangement, a third party owns equipment and then leases it to an organisation at a given rate for a set period of time. At the end of the lease, the organisation has the option to purchase the equipment, continue leasing, return the equipment or lease new items.

In the UK and elsewhere, this option has been favoured for photovoltaic projects. In a typical example, photovoltaic panels would be installed and leased by a third party or parties to a building owner over an agreed period of time. During that time it is anticipated that energy savings (and payments such as feed-in tariffs) will cover the cost of the lease.

Leasing and lease-purchase agreements provide a means to reduce or avoid the upfront capital investment for energy efficiency improvements. Equipment lease financing can work for equipment ranging from lighting all the way up to major heating and cooling systems. Benefits include the fact that there is no capital investment on the part of the building owner and the low entry cost of acquiring new equipment.

Since a lease often does not require a down payment, it is the equivalent of 100% financing for an energy efficiency retrofit project. Most lease periods range from five to ten years.

Government financing

In the UK, there is a new player in commercial energy efficiency retrofit: Central Government. In the past three years there have been several newly created programmes designed to encourage energy efficiency retrofits. These include the Feed-in Tariff (FiT), the Renewable Heat Incentive (RHI) and the Green Deal.

Under the FiT scheme, the Government provides payments for electricity generated from renewable energy technologies. The payments are for every unit of renewable electricity (kWh) generated, with one price for electricity consumed on-site and another price for excess energy exported to the grid. The amount of the tariff unit price depends on the size and type of the technology installed.

While there was much excitement at the launch of the FiT scheme, enthusiasm has waned in the commercial market. This is due to changes in how the programme is administered (primarily because of the decision by Government to drop tariff levels). A series of changes in rates, plus concern about returns on investment, has meant that many planned schemes for renewable energy retrofits have stalled.

The RHI for non-domestic buildings was rolled out in November 2011. The scheme operates similarly to the FiT but is for renewable heat, not electricity, generation, such as that provided by biomass or solar thermal. Commercial interest in the RHI has been low. This is, in part, due to unfamiliarity with the technologies covered, but is also in response to discouraging experiences with the FiT.

The Green Deal, launched in 2013, is a UK Government initiative that provides loans for energy efficiency solutions, which will be paid back via the electricity bill. Acceptable solutions are those that follow the 'Golden Rule', where the monthly savings in the energy bill from the retrofit are equal to or greater than the monthly cost of paying back the loan over a standard loan period (typically 25 years). The loan stays with the property and the new owners will continue to repay the loan as they will also benefit from the reduced energy bills.

The Green Deal is by far the most ambitious Government initiative to date to spur energy-efficient retrofits of commercial properties. But, in the commercial market, uptake of the programme has, so far, been limited and is likely to continue to be so. This is due, in part, to confusion about the policy but also because the incentives have not been sufficiently tempting to entice businesses to act. The Green Deal has been a particular disappointment, and far fewer applications for funding have been made than were envisioned.

CHANGE

Despite the lacklustre embrace of traditional and Government finance models, there is significant change under way. In the past five years, new regulation and changing market dynamics have brought energy efficiency to the top of the agenda. Whereas traditional arguments for energy efficiency used to focus (unsuccessfully) on operational costs and savings, the newer paradigm is obsolescence and long-term value. Framed in this way, the stakes are much higher. Owners, occupiers and lenders are adopting new attitudes to energy efficiency.

The energy efficiency of a property is rapidly becoming a crucial issue for the property owner and occupier market. This is not so much because energy costs are rising (they have doubled in a decade) but because the market is beginning to financially reward environmental performance. In the past five

years, the amount of information that is available about buildings and their performance has increased significantly. Alongside this, occupiers, particularly corporate occupiers, have been put under pressure by their stakeholders to demonstrate sustainability credentials. The market has witnessed overarching institutional changes, such as greater property availability and shorter lease lengths. Occupiers now know more, want more and have more choice.

Regulation

It is worth remembering that most of the sustainability regulation that now affects property has been enacted in the past five years. Up until 2009, it was difficult or impossible to know a property's energy efficiency. It is still not legally mandatory in the UK to require commercial properties to disclose actual energy performance, although that step has been taken elsewhere in the world with ramifications for retrofit financing. This has not stopped UK occupiers from asking for data on the actual energy performance of buildings – in fact, it is a regular feature of most new due diligence.

Beginning with Energy Performance Certificates (EPCs) through to mandatory carbon reporting, there has been a steady trickle of legislation that impacts on transactions. Although EPCs have been criticised generally as poor predictors of actual performance, they have nevertheless spurred retrofits.[7] They will continue to do so as companies run up against the threat of being unable to rent or sell energy-inefficient properties in 2018. Despite their long-standing reluctance, landlords will undertake retrofits and allow their tenants to benefit because they stand to lose more money if they do nothing at all.

The amount of stock affected is not small. An estimated 18% of non-domestic stock would currently be restricted from transactions unless it was made more energy efficient.[8] Of this 18%, nearly two-thirds is in the commercial real estate sector. Landlords are already reviewing existing stock to introduce energy-efficient retrofits – if for no other reason than to keep their assets viable.

Mandatory carbon reporting, recently introduced in the UK, is also likely to impact on an owner's decision to undertake energy-efficient improvements. Where disclosure has occurred, for example in the USA and France, the reaction has been the same. Tenants who were previously uninterested in energy or carbon are suddenly galvanised into action when they can judge the performance of their premises relative to buildings around them. Companies can use this information in the selection of their premises or during a rent review to argue for lower rates.

In the UK there has been a clear and convincing trend of regulation to encourage disclosure and transparency around the issue of energy efficiency. Although the market for energy-efficient property is still new, there is little doubt that building energy performance will guide future occupier choices. Properties that fail to perform environmentally will fail to perform financially.

The market

While regulation has helped to build a market for energy-efficient buildings, the market has also moved on its own. In the past ten years, the average length of a UK lease has fallen dramatically, while the proportion of five-year leases has doubled. At the same time, the amount of existing stock available has risen significantly. This means that occupiers have more choice and more chances to exercise that choice. Owners wishing to attract and retain tenants have to be much more attuned to their preferences than the previous generation of landlords.

Under these changing circumstances, owners are recognising that retrofits will be a significant factor in retaining and attracting new occupiers. Offering occupiers a building that is cheaper to run and that demonstrates top sustainability credentials makes good business sense.

The business case for sustainability is beginning to be made, not in terms of costs and payback, but in terms of value and obsolescence. Landlords are less concerned with whether tenants reap the savings from retrofits than whether they will stay and continue to pay rent once improvements have been added. Retrofits have moved from a cost to be avoided to a cost to be incurred because energy-efficient properties now have a value that they previously did not have.

ENERGY EFFICIENCY AND VALUE

One of the most promising areas of research that will inevitably spur interest in retrofits is the connection between energy efficiency and asset value. It is only in the past five years that practitioners have been able to establish a connection between the energy performance and the financial performance of buildings. This emerging line of enquiry and the relationships it has begun to establish have transformed the perception of the importance of energy efficiency in the marketplace. Indeed, essentially all of the new interest in retrofits and creating funding vehicles stems from the realisation that retrofits are less about cost savings and more about asset value and protection.

It is beyond the scope of this essay to set out the findings of academic and corporate research on energy efficiency and market value. It is also worth noting that these studies have been the subject of intense debate and criticism about methodologies and findings.[9]

Having said that, three important findings from the research must be listed here:

1. no matter what measure of energy efficiency is applied (design or operational), the financial performance of buildings is better when those buildings are efficient;
2. while the magnitude of increased value from energy efficiency in properties varies, the direction of the trend does not; and
3. this is true across countries and across sectors.

This so-called 'green premium' exists for other aspects of sustainability more broadly (such as having a building certification such as BREEAM or LEED), but the relationship between energy performance (in operation) and asset performance appears to be the strongest.

This means that the actual energy performance of buildings (as opposed to design features) is becoming a critical factor in real estate decision-making. Indeed, studies by real estate professionals show that preferences for energy-efficient buildings are clearly on the rise and that many organisations will not occupy poorly performing buildings.[10]

Even though full transparency in energy performance is not yet mandatory in the UK, this has not prevented the market from taking its own course. As a standard part of due diligence, real estate teams now require evidence of past energy performance. Energy performance is emerging in discussions at rent reviews, where companies are requesting lower rents for poorly performing buildings. In other words, where tenants once looked the other way, they are now focusing intently on the energy performance of their premises.

This is partly due to concerns about the rising costs of energy, but for companies there is an issue of much greater value at stake: reputation. It is becoming impossible for corporate occupiers with highly visible corporate sustainability requirements to occupy poorly performing buildings. For an average service company, 90% of reportable carbon emissions stem from building occupation. As integrated and mandatory carbon reporting becomes the norm, corporate occupiers are re-examining their premises and energy efficiency is having an impact it never had before.

Obsolescence

It is the fear of lost value that will drive retrofits in the future. A recent British Council for Offices (BCO) report notes that there is an emerging industry belief that assets with poor energy efficiency will see significant depreciation in the future.[11] This means that owners need to develop a more proactive approach to retrofits and to provide current and prospective tenants with spaces that satisfy their requirements.

This same report finds that developers, investors and occupiers all rate energy efficiency as either the first or second most important driver of obsolescence over the next five years. Buildings that fail to undergo retrofits will see their value eroded much faster than ever before.

OTHER FACTORS

Buildings lose value over time – that is one of the first rules of property. What is more alarming is how quickly value can be lost in a market that has an increasing amount of choice. One of the defining characteristics of the commercial market in the past ten years has been the growth of 'exit' – that is, the ability of occupiers to review property decisions and leave buildings that no longer meet their requirements. Whereas owners could once turn a deaf ear to tenant concerns, secure in the knowledge that they had a long lease in place, this is no longer true. Tenants now have the upper hand. The fact that they increasingly cite energy concerns as their top priority is forcing even reluctant owners to make changes or risk losing tenants. Losing a tenant incurs a much greater cost than upgrading the building and this too has enhanced the business case for investing capital in energy efficiency improvements.

Lease length

In 2000, leases of five years or less represented just 10% of the commercial market. Today, this number is closer to 50%. Conversely, the proportion of leases of 20 years or more has declined from 25% to 5% over the same period (see Figure 3.1).

It is not difficult to see what is happening – the secure long-term leases which gave owners the advantage, and which may have encouraged complacency, have disappeared. Tenants now have more opportunity to assess their assets

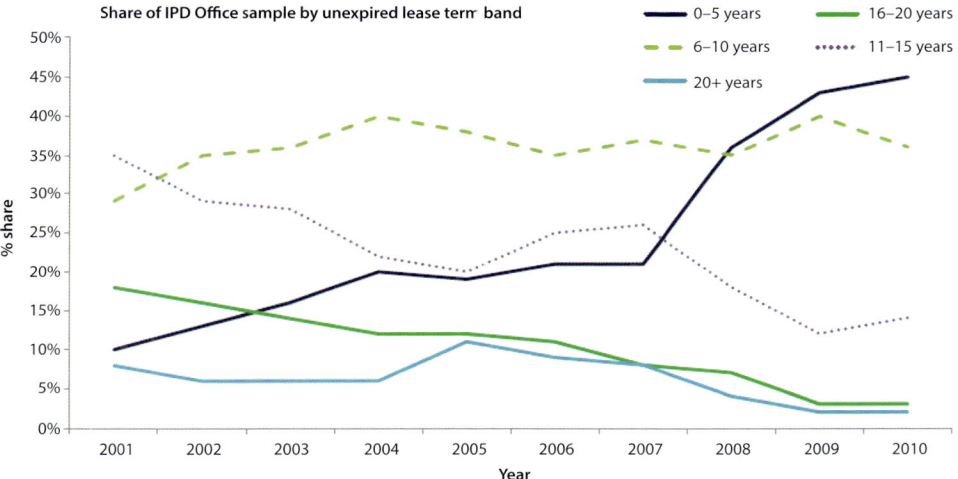

Share of IPD Office sample by unexpired lease term band

Legend: 0–5 years · 6–10 years · 11–15 years · 16–20 years · 20+ years

% share — Year

3.1 Lease Length, 2000-2010

and consider whether their physical assets are enhancing their corporate image and finances. Certainly, energy efficiency will not be the only reason why a tenant chooses or leaves a property, but it is, increasingly, a bargaining chip and present owners must be far more attuned to what their tenants demand, regardless of who benefits from the operational savings of energy efficiency.

Availability

One of the most significant changes in the commercial property sector over the past ten years has been the surplus of existing property. Much of this property is sub-par in terms of energy performance. And so, while there are questions about how deeply energy efficiency preferences have penetrated into the market, it is clear that existing properties looking to attract tenants will have to try harder.

In 2000, the take-up rate and the availability of existing property were roughly equivalent. Presently, availability exceeds take-up by a factor of four. So, at a time when occupiers have more exit opportunities, they also have far more choice. Because there can be such a vast difference in energy performance between properties that are only a few years apart in age, energy performance is bound to be one of the determining elements in the choice between existing properties. In a tighter market, energy concerns may be of less importance, but with plenty of choice available occupiers who have a preference for energy-efficient properties will inevitably seek them out. This means that the threat of obsolescence of poorly performing properties is only increasing.

Energy costs

It is often noted that energy costs comprise only a small proportion of a company's operational costs. This is true, and historically occupiers have focused primarily on rent and not the costs of operation in property selection (see Figure 3.2).

However, as money has become scarcer and the true costs of poorly performing buildings better known, energy costs have become a consideration. This is true not only because the rises in energy costs have been sharp and show no sign of abating, but also because there are now carbon costs associated with building operation. The costs related to energy and carbon are, admittedly, small compared to total costs, but they are increasing, focusing attention on operational costs in greater detail.

During this time of greater exit and choice (from the year 2000 to the present), electricity and gas prices have more than doubled.

No other operational cost has undergone anything like this rate of inflation. Moreover, because Building Regulations have compelled such a rapid increase in energy performance, it is often the case that two properties which are close in age can have widely differing energy usage.

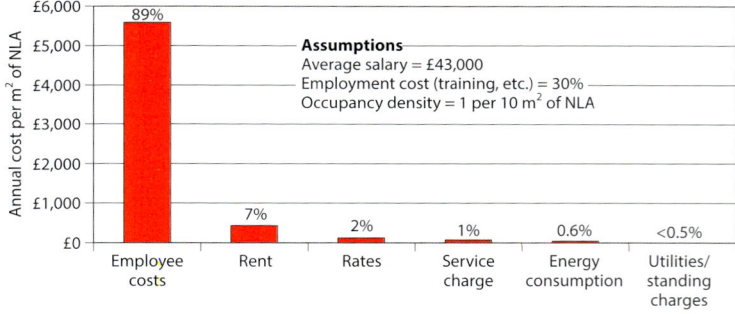

3.2 Typical business costs of an office building in London
(NLA = Net Lettable Area)

This raises two very interesting issues. First, the astute occupier can lease a property that is akin to another in finish, location and general age, but which has a very different energy profile. Second, the difference in that profile will be due, not to occupier activities, but to physical characteristics that are the responsibility of the landlord. Put slightly differently, there can be dramatic differences in the performance of buildings that are only a few years apart in age and these differences will be due to the building fabric and/or systems (which is the landlord's responsibility) and not to the tenant's use.

For example, a property in London built after 2006 can have operational costs that are £3 per square foot per year lower than a building just a few years older used in exactly the same way. In this type of scenario, overall energy costs would drop from £8 per square foot to £5 per square foot, almost a 40% reduction. Controlling for tenant activities, older buildings are generally more expensive to operate. The market is beginning to appreciate this and landlords will have to demonstrate that they are making improvements to existing stock if they wish to compete with newer properties.

There is no denying that energy costs will increase – it is merely a question of by how much and how soon. Estimates from the Department of Energy and Climate Change suggest that energy prices will double, not in ten years but within five. This means that the costs of owning poorly performing buildings will only be further magnified in the future.

INTANGIBLES

It is worth remembering that carbon emissions were once regarded as an intangible – difficult to define, understand, measure and value. It is no wonder that it was routinely underestimated how carbon performance would affect the value of many buildings, even some of those built within the past decade. It was easy for developers, investors and occupiers to ignore advice about the importance of carbon performance, because professional bodies themselves were unprepared to speculate on the value implications of an emerging 'soft' topic like carbon. Yet, if the last few years tell us anything, it is that soft topics can have hard edges and the value of intangibles must be anticipated.

The monetary value of energy-efficient properties is becoming important (both in terms of reduced costs and increased value), but this is only part of the story. The *reputational* value of owning or occupying energy-efficient properties is difficult to quantify but is certainly destined to feature in transactions in the future. Because buildings increasingly reflect the values of those within them, it will be difficult for companies that wish to project the right brand image to occupy buildings with poor energy and carbon performance.

It is a recent, but widely held, assumption that a prime building is a sustainable building. Relatedly, there is an emerging consensus that a sustainable organisation is the one likely to attract and retain the top talent. One of the unusual (but upon reflection unsurprising) developments in sustainability has been the recent involvement of human resource professionals in property decision-making. They recognise that projecting a desirable image to potential recruits and employees is a critical financial decision for their organisation, with economic stakes that far exceed the cost of energy. It is when sustainable performance (which is largely defined by energy use) becomes associated with something far more important – image, mission and morale – that the business case for energy-efficient, sustainable properties becomes evident. To think that energy efficiency is just about the building is to miss a giant opportunity.

Using buildings to project a desirable corporate image is hardly new. What is new is the corporate image that is now considered desirable. Recent reports on upcoming workers (of the kind that most organisations want to attract) find that traditional commercial offices hold little attraction, primarily because they are regarded as unsustainable.[12] The financial and environmental are converging in ways that were unthinkable to the property industry a decade ago and have yet to be fully appreciated. As a concept, sustainability is widening and narrowing at the same time, as occupiers consider all the ways that buildings can add value to their organisations while simultaneously zeroing in on what specifically works to improve their profile. One of the ways to judge whether a building 'works' in terms of sustainability is energy in use. Buildings that fail to meet this criterion will almost certainly be deemed unsustainable, no matter what other attributes they may possess. Energy efficiency is not all there is to sustainability, but it is the litmus test and the key to capturing intangibles of significant worth to companies.

Value and values are inextricably linked and, in addition to quality, sustainability is being linked with certain other concepts. These include 'modern', 'disciplined', 'efficient', 'committed' and 'trustworthy'. Likewise, smart occupiers are beginning to use buildings as assets in the true sense of the word, using property performance to build engagement, loyalty, purpose and pride. Conventional, unsustainable buildings hold little attraction for the next generation of workers, who have been raised with a keen awareness of sustainability.

WHERE IS THIS GOING?

Bringing older stock up to new standards may well be virtually impossible, financially and physically, in many instances. It can be difficult enough to encourage owners to undertake light-touch energy improvements, never mind the extensive deep retrofits that many commercial buildings require. Take away the intervention of a major transformation in business as usual, and improving building energy performance may well continue to be an interesting theoretical discussion that goes nowhere commercially for much of the existing stock.

But there are signs that a fundamental shift is under way. Studies of occupiers note that sustainability is 'all they want to talk about'.[13] Sustainability issues have moved from a periphery issue into the mainstream of property decisions, particularly as investors and lenders consider the longer term profitability of their portfolios. Naturally, most of this discussion centres on new prime assets and top corporate occupiers, but a general trickle-down effect can be seen and should be expected. This is why even reluctant owners have begun to undertake light-touch refurbishments in secondary assets that would have been unlikely just a few years ago.

Professional real estate observers note that deep retrofits will inevitably follow as regulation and market drivers make poor energy performance more unacceptable and new investment vehicles make more aggressive interventions less untenable. Clearly, the rise in both the number and type of financing arrangements is indicative of a general interest in deep retrofits, since difficult projects are the very reason behind the formulation of special purpose vehicles.

BUILDING IS NOT THE SOLUTION

There is no doubt that many assets have poor energy performance because they are poorly built or have physical characteristics that raise energy use. For these types of assets, deep retrofits will be necessary and some type of construction activity is a must. These are the difficult cases and there are many of them. But to think that building – even building of the best kind – is the answer to the carbon problem is a mistake. It displays the same kind of thinking that got us into the problem in the first place.

Most buildings perform poorly because they lack the resource that is in shortest supply – diligence. Management and commitment are just as important, if not more important, than physical features and technology. A recent study of low carbon fit-outs found that the most significant savings were captured without capital expenditures at all, but instead reflected a commitment to understanding and implementing ideas that work.[14]

Better monitoring and verification of actual energy use are vital to demonstrate to executives the business case for improvements and to evaluate what works and what does not for implementation in the next project. The simple, effective, cheap actions – education, monitoring, management, collaboration – are both highly effective and severely lacking in today's buildings. There is, in the current market, a clear shift away from what we put into buildings to what we get out them. No amount of technology or good design will ever supplant the need for human diligence, care and performance.

CONCLUSION

The building industry is becoming infinitely more complex, from the multi-disciplinary approach to building to the multifaceted expectations of occupiers. Of the three barriers to energy retrofits, the most significant one – attitudinal – is also the one undergoing the most dramatic change. The reluctance to undertake energy retrofits has never been primarily about costs and savings or even the technologies themselves. Instead, it has been about the perception of the role that sustainability plays in property decisions.

That role is changing significantly, and faster than we imagined possible. Energy efficiency is about sustainability, but it is also about something far more important – image, brand, value and quality. These issues matter to occupiers and landlords are beginning to improve energy efficiency in order to boost their overall financial performance and viability. Expectations and attitudes about energy efficiency have, in a few short years, changed from indifference to indispensable. It is not about cost but about value, and environmental and financial performance are becoming indistinguishable.

How far energy efficiency will (and can) be taken for the poorest of existing stock remains unclear but, for many properties, the ability to remain as viable assets will be synonymous with the ability to achieve acceptable levels of energy performance. New investment vehicles are emerging to reduce the financial and technical risk and these will undoubtedly make investment risks less daunting to take. However clever and innovative the investment vehicle may be, it is important to remember that what propels these vehicles is a fundamental shift in the perception of energy use and what it means. From the owner to the occupier to the energy provider, there are clear profits to be made from using less energy. The traditional split incentives and competing interests are diminishing and in their place is an emerging consensus about building performance and its place in overall business success.

ENDNOTES

1   Better Buildings Partnership, *Low Carbon
     Retrofit Toolkit: A Roadmap to Success*
     (2010), p. 6.

2   *Ibid.*, p. 4.

3   Technology Strategy Board, *The
     Retrofit for the Future Projects: Data
     Analysis Report* (2013) and Institute for
     Sustainability and Arup, *Delivering and
     Funding Housing Retrofit: A Report of
     Community Models* (2013). Also see
     R. Francis and W. Wright, *Counting
     Carbon, Counting Costs: Achieving
     Performance in Retail Fit Outs*, British
     Council of Shopping Centres (2012).

4   For a good discussion of developing
     standards, see A. Buonicore, 'Energy
     Efficiency Retrofit Options for the
     Commercial Real Estate Market', *Building
     Energy Performance Assessment News*
     (2012).

5   See M. Brack, 'Energy Retrofit of London's
     Commercial Building Stock', *worldcities.
     org* (11 February 2013).

6   For a discussion of light versus deep
     retrofits see Chapter 1 by Sunand Prasad.

7   For a discussion of EPCs as predictors
     of energy performance, see Jones Lang
     Lasalle and Better Buildings Partnership,
     *A Tale of Two Buildings: Are EPCs a True
     Indicator of Energy Efficiency?* (2012).

8   Royal Institution of Chartered Surveyors,
     Property in the Economy 2012 (October,
     2012), p. 32.

9   For a good review of this literature, see
     World Green Building Council, *The
     Business Case for Green Building:
     A Review of the Costs and Benefits for
     Developers, Investors and Occupants*
     (2013), pp. 110–133.

10   Deloitte, *UK Real Estate Predictions 2013:
     The Evolving Market* (2013).

11   British Council for Offices, *Change for
     the Good: Identifying Opportunities from
     Obsolescence* (2012).

12   PwC and Urban Land Institute, *Emerging
     Trends in Real Estate: Europe* (2013).

13   *Ibid.*, p. 12.

14   R. Francis and W. Wright, *Counting
     Carbon, Counting Costs: Achieving
     Performance in Retail Fit Outs*, British
     Council of Shopping Centres (2012).

chapter 4

RODERIC BUNN

FROM POST-MORTEM TO LIFE SUPPORT

BUILDING PERFORMANCE EVALUATION AS A DESIGN TOOL

INTRODUCTION

Building performance evaluation (BPE) originated from post-occupancy evaluation (POE), providing owners and designers with data on a range of factors, such as energy use, comfort and user satisfaction. BPE embodies a much broader application of feedback study, with the results specifically used to inform and improve design decisions. In retrofit projects, which already provide a 'baseline' model, this approach can be particularly effective provided that a pragmatic, common-sense approach is adopted; the study of building performance is as much an art as a science.

Implicit in the terms 'retrofit' and 'refurbishment' is a belief that the process will lead to a better performing building. Whether in terms of greater fitness for purpose, extended life or energy efficiency, the expectation is not unreasonable. It does depend, of course, on the state of the building prior to the refurbishment, the degree to which the building will be altered or improved, and whether the refurbishment leads to a change of use.

Take two examples that appear to be polar extremes: the redundant rural Victorian primary school and the crumbling 1960s inner-city office block – period-defining buildings that conjure crystal-clear images in the minds of the reader. The first, a pitch-roofed, solid-walled, bay-windowed building of distinctive period character. The second a grim, concrete, multi-storey, heavily glazed, flat-roofed, water-stained blot on the cityscape. The former will be loved for its longevity; the latter despised for the same reason.

Both can be refurbished to the latest standards and put to new uses: the school converted into a dwelling, the office block into flats or reinvented as a contemporary office. The end-user's expectations for both, however, will be very different. The prospective occupier of the old school will want the virtues of the original building to be retained. This may require some degree of compromise over things like circulation, room sizes and ceiling heights, and possibly energy consumption if improving insulation or airtightness is either technically difficult or falls foul of heritage requirements. Nevertheless, despite the change of use and what could involve extensive internal remodelling, the performance objectives will be tempered by the new owner's desire to retain the school's original character. Shortcomings will be considered tolerable quirks. Ergonomic problems or space heating deficiencies will be forgiven, accepted as the penalty for living in an iconic building.

The refurbished office, on the other hand, will have few virtues worth retaining. Irrespective of whether the refurbishment involves mere replacement of the windows or a wholesale strip back to the structure, no one – neither the project funder, the landlord, the maintenance team nor the future occupants – will tolerate the retention of the original building's limitations. No forgiveness factors will be at work. Furthermore, the success of the refurbishment will be judged solely by contemporary norms. Expectations will demand that the refurbished building possesses a performance equal to, or better than, a new office building. Any inherited physical constraints simply won't count.

This comparison between two typical refurbishments may be simplistic, but illustrates a vital point: the performance expectations of refurbished and retrofitted buildings depends almost wholly on the local context. Success purely depends on whether the clients and end-users get what they are expecting.

The issues at stake are whether these expectations are reasonable, given the circumstances. Are they well expressed? Are the developer and the design team fully aware of the constraints that they are aiming to overcome? Can any performance outcomes be defined, relevant metrics set, and ways found to measure them when the building is brought back into use? These are difficult questions, but answering them is vital if the building's refurbishment is to be judged a success.

Success encompasses a wide range of factors: scientific, such as heat loss and levels of airtightness; quantitative, such as energy consumption, carbon emissions and indoor air quality; statistical in terms of comfort perception; and qualitative, such as whether occupants perceive that their needs are met.

Such factors – and there are many others – define the building's operational performance. When the operational outcomes have been defined, it is then a matter of the project team designing, constructing and commissioning the building to achieve those outcomes and, once the building has been handed over, to remain engaged and use methods of building performance evaluation to check whether they have been achieved. Success should certainly be celebrated, but any shortfalls must be investigated and, where possible, interventions made to put them right. This is the only way in which the open loop between design intention and operational performance can be closed.

THE ORIGINS OF BUILDING PERFORMANCE EVALUATION

Interest in building performance evaluation is a fairly recent phenomenon. After all, there has been no legislative imperative to do it. Even in 2013, Building Regulations still presumed that compliance with energy standards can be both proved and achieved at the early stages of design. This bolsters the presumption that design quality – a nebulous concept if ever there was one – is well able to survive the stormy waters of construction, unscathed by cost-cutting, rushed commissioning and perfunctory handover.

Such is the ingrained belief that building performance can be defined at design, that the design professions and their support structures – schools of the built environment and the professional institutions – never twigged that a performance gap could exist, let alone be their responsibility to close. As a consequence, professional terms of appointment, forms of contract and payment regimes are all geared to the design, delivery and handover of a built asset, not to custody of its operational performance. It is true that project team members are obliged to resolve snags and correct defects, but this has very little to do with performance outcomes.

Historically, all parties involved in a building's procurement have been blind to the performance of the thing they are paid to deliver. No one in the construction supply chain is conditioned to remain engaged after handover. After all, nobody would willingly volunteer to face the inevitable post-handover questions. But, as former Atkins chief executive Keith Clarke put it, 'Construction professionals have the privilege and honour of being obliged to think'.

So, what has changed to force the professions to think? Essentially, it boils down to one word: energy. The use of it, the waste of it and the cost of it. Cost has always been an issue, but compared with the costs of procuring a building and of employing people within it, energy has been a long way down the list of priorities. Now, thanks to global warming and consequential climate change, wasteful energy consumption comes with more worrying baggage:

generating damaging carbon dioxide emissions that international legislation, backed by public pressure, is seeking to cut.

Environmental rating schemes that promote energy-efficient technologies and reward the use of on-site renewables are now common currency. Most clients want – nay, expect – greener buildings. When mandatory energy performance certification for public buildings was launched in 2008, clients were not sufficiently well-informed to question whether the ambitious predictions of energy consumption made at the design stage would be achieved in practice. And, because the construction professions were ignorant of the difference between performance at design and performance in practice, and knew little about the complex interactions between systems and factors such as out-of-hours running and the consumption from plug-in power devices, they were not in a position to disabuse their clients of the notion that design could solve everything.

The scene was set for a rather rude awakening. For when the first Display Energy Certificates (DECs) went up in receptions across the country, the yawning gap between design ambition and operational reality became clear. Buildings designed to a B or C energy rating were only achieving operational ratings of D and E. Regardless of the fact that DECs are not truly comparable with design energy certificates, all that mattered to clients was that their buildings were not as green as they had been led to believe. Clients and their management and maintenance teams struggled to understand what they had been given, such was the unexpected complexity of the engineering systems and the computerised controls.

The warnings from history were there for anyone who cared to check. Post-occupancy evaluations (POE) from the 1970s onwards had already found performance problems with ostensibly low energy buildings.[1] However, the work was mainly conducted by academic researchers. The construction industry remained largely ignorant.

It took a ground-breaking series of post-occupancy investigations, carried out and published by the UK journal of the Chartered Institution of Building Services Engineers (CIBSE) for the problems to be widely recognised. The PROBE series of investigations spanned six years, from 1995 to 2001, and involved the forensic analysis of 20 notable buildings that had been publicly celebrated for their green credentials at the time of their completion two years before.[2]

When put under the research microscope, the often award-winning buildings invariably failed to live up to their design aspirations. The few successes were vastly outnumbered by buildings with energy performance and occupant satisfaction that was mediocre at best. Few came anywhere near the best practice energy benchmarks, let alone matched them or beat them. It was not just on energy that the building were underperforming – there were many failings in their manageability and usability. In numerous cases the building operators where overwhelmed by the demands of overly complex systems that the users were neither expecting nor trained to manage.

The PROBE project was not only notable for lifting the lid on the truth of building performance, it also popularised a practical, robust and comparatively rapid process for analysing building performance. It also helped to develop a set of tools that have become the de facto standards for undertaking POE.

A key virtue of PROBE was its objectivity and avoidance of any vested interests that might have an influence on the direction of the research and its reporting. The original designers were deliberately kept at arm's length while the research was conducted and only asked to comment once the results were available. This proved to be vital. The construction professions, particularly, found it hard to have dirty washing aired in public. They had too much to lose.

Five years after PROBE, the Carbon Trust funded a larger set of building studies in two parallel research projects – the Low Carbon Building Performance and the Low Carbon Building Accelerator programmes.

These four-year projects, where research teams followed projects from their inception into the first year of operation, were primarily concerned with promoting low and zero carbon technologies, but included some post-occupation analysis.[3]

While some buildings in the programme performed close to their expectations, just as in PROBE most did not. Their ostensibly low carbon technologies often proved to be fragile and delivered rather less than was expected. With hindsight, perhaps too much faith was invested in their performance potential, and not nearly enough attention paid to all the other things that have to be right: sensible budgeting, good design detailing, installation quality and thorough commissioning, for example. The high hopes and honest intentions of green innovation plunged down the crevasses of a disjointed procurement.

In 2010, the UK Government became sufficiently concerned to invest £8 million in studying the building performance problem. Only now the focus was not on POE; the research had grander ambitions. It sought to study buildings under construction and to follow through in the first two years of operation. It also wanted to find ways of closing the performance gap so that it would not undermine its low and zero carbon targets, and the mechanisms to achieve them, such as the Green Deal.

What the Government's innovation agency, the Technology Strategy Board, needed were some research methodologies, and for these they turned to PROBE for inspiration.

TOOLS OF BUILDING PERFORMANCE EVALUATION

PROBE used three barometers to assess a building's performance: an energy survey, an occupant satisfaction survey and a walkthrough of the building and analysis of its engineering services. The energy survey disaggregated the building's annual energy consumption by its various end-uses, including the client's plug-in electrical power loads and other sources of consumption, and reported the breakdown in kilowatts and kilograms of carbon dioxide per square metre of treated floor area. The occupant survey was based on a simple paper-based questionnaire. This captured statistically valid data and anecdotes about the building's performance from the permanent occupiers of the building.

The results of these two studies were cross-referenced and compared to the findings from a forensic walkthrough conducted by experienced building analysts with a background in architecture, engineering and building physics. Snapshot measurements of lighting levels, space temperatures and control set-points were gathered, along with any physical measurements, such as airtightness test values.

The resulting triangulation of results left very few questions unanswered except, in many cases, the root causes of why the building was designed and built the way it was. As the project teams had long since disbanded, the PROBE research team only had the intentions and recollections of the building owner, the architect and the services engineers to go by, recollections somewhat compromised by contractual obligations. The input from the builder, key sub-contractors and the commissioning engineers was largely lost to history.

The tools of assessment currently used in PROBE didn't exist at its inception. Necessity, however, is a surprisingly fertile mother of invention. An energy spreadsheet tool for offices was gathering dust on a researcher's shelf. This was the Energy Assessment and Reporting Methodology (EARM), subsequently published by the CIBSE as *Technical Memorandum 22* (*TM22*). Similarly, an occupant survey was created by adapting an office environment survey devised by social scientist Adrian Leaman to investigate sick building syndrome (a hot topic during the early 1990s). This was revised and streamlined into a structured questionnaire which could be rapidly administered and analysed, and the results compared with a benchmark dataset. This became the Building Use Studies (BUS) methodology (now owned by Arup and available through a network of resellers).

THE GRADUATED RESPONSE TO INVESTIGATION

As well as establishing the key tools to perform POE, PROBE did something equally useful. It determined that understanding building performance did not require the gathering of large complicated datasets. Robust investigation could be done quickly and cost effectively. Yes, there could be complex interactions between systems that took time and troubleshooting skills to unravel, but in most cases the causes of underperformance were easy to spot.

If the reader takes away just one lesson from PROBE, it is the importance of a graduated response to understanding building performance: start with things that are simple and easy to study, and don't get bogged down in the detail too quickly. One can graduate to more detailed analysis (of energy data, for example) if early study fails to identify the problem.

Corroboratory evidence from thermography and airtightness tests can identify areas of leakage and cold bridging for example, but such tests would not be necessary unless initial evidence points to a lurking problem. For example, if the heating system is well controlled and gas consumption is reasonable against prevailing benchmarks and the design estimations, and correlates within reason to annual degree–day data, why analyse it any further? But, even if heating energy consumption seems wayward, that is still not the trigger to start lifting floorboards. It is best to build up knowledge with non-invasive analysis: investigate the control settings first and do some local checking of air leakage using hand-held smoke or puffer pencils. If that does not answer the questions, you might start to dig deeper and see if heating and refrigeration systems are fighting each other. Similarly, if lighting energy consumption is high, check the installed wattage of fittings and carry out some local checks on lighting levels using a lux meter rather than gathering gigabytes of sub-meter data and wasting time and money carrying out lengthy data analysis.

As Bill Bordass says in Chapter 2, keep things simple. The rule applies to the analysis of buildings just as much as it does to their design. Usually, the reasons for underperformance do not lie far below the surface. Only get out the big research guns when you can't get to the root causes quickly and easily.

FROM POST-MORTEM TO LIFE SUPPORT

While the post-occupancy analysis of buildings is important in understanding what works well and what does not, it will not help the design professions to do the job unless the lessons learned are captured, retained and used to inform the design of the next building. But, even then, the design professions will still be one building behind on the learning curve. A far better strategy is to try to understand more about the specific building in hand before the client freezes their requirements and, similarly, before the design brief is written. In that regard, retrofit scores hugely over new build: the building and its occupants already exist and are, therefore, available for study.

One of the great virtues of the PROBE tools, and others that have been developed since, is that they can be used as pre-design tools as well as post-occupancy analysis tools. An energy spreadsheet created pre-retrofit can enable the mapping of energy consumption by end-use. Current inefficiencies in the building can be identified, and this can help the client and design team to focus on the areas that require improvement.

A greater understanding of the building's zones and end-uses will also help the designers to devise a suitable electrical sub-metering strategy. It should certainly help to prevent the widespread practice of putting sub-meters on every available electrical circuit and end-use, a policy which can lead to over-metering that no one in the facilities team will ever have time to analyse. Having too many meters is as unhelpful as having too few.

Having said that, energy surveys of existing buildings are often more an art than a science. It is not unusual for an old building to have limited metering. It might have oil-fired boilers with consumption records based on deliveries and vague tank gauges. It may have other oddities, such as later extensions served by local air-conditioning units. Its hours of occupation may be complicated,

and it may even have more than one use. The author has been called upon to carry out an energy survey of a primary school only to find that it had a church within its walls used by the local community both within and outside of normal school hours. Separating out who uses what part of the building and when requires attention to detail and the careful recording of assumptions.

The lesson here is that even with rigorous analysis you won't always get to the truth. It is better to settle for getting the best obtainable version of the truth and accept gaps and estimations. In most cases that that will be a good starting point.

INVOLVING THE HUMAN

If you think measuring energy is difficult, try understanding people. Set up properly, an energy meter will largely do what it is told. Flow devices like gas meters can be a bit wayward, and heat meters even more so. But, generally speaking, these devices are reliable once calibrated and correction factors are applied to balance out any errors in measurement. People, on the other hand, are impossible to calibrate. They are volatile; there is a high level of randomness in a given population, and there are often gender differences, particularly relating to what constitutes comfort conditions. However, in a retrofit context, if you want to know the needs and expectations of the occupants, and to design something that meets those expectations, then it pays to examine people's needs and wants and use the knowledge to inform the design strategy.

There are many ways to survey people. Semi-structured interviews with individuals, such as building managers and other stakeholders, will usually throw up some useful insights, but bear in mind that the views of senior managers, for example, will be very different to those of general occupants. This is not simply a matter of hierarchy. People whose jobs tie them to their desks will have different sensitivities to those with the freedom to roam and who are therefore able to escape any discomfort problems. To managers, many usability problems may appear trivial or low priority. To other users, a number of small issues can add up to a significant and undesirable situation.

Group meetings with occupants can be enlightening, although you should not be surprised if wildly contrary views are expressed. In any group there will be leaders and followers – those with strong views will express them, influencing the group dynamic, while others will say little or remain silent. Many will go along quietly with the consensus. In any case, people who are content or happy with the building's conditions probably won't turn up to these meetings. These factors will skew the results, which is dangerous if those holding the meetings are seeking to have their own beliefs corroborated or their prejudices confirmed.

For these reasons (and many others) group meetings can lead to a biased report of the building. The outputs will also be purely anecdotal unless people are asked to fill in some kind of scored questionnaire as well. However, questionnaires used in group forums are prone to pressure from the researcher, cross-referencing within the group and oversight by managers. All these things introduce distortions.

Internet-based surveys can be used, but this is still not ideal, as conditions might be different in the building during the period in which a survey is carried out. One cannot control how the survey forms are filled in – a group huddled around their computer screens swapping notes introduces bias in the same way as a group collaborating over their paper questionnaires.

Nothing beats a paper-based questionnaire filled in on a given day by all the individuals in the building on that day. With well-crafted questions framed in the right way, and a scoring system that is simple to code and analyse later, a survey can generate some useful statistics.

Before embarking on creating a questionnaire from scratch, bear in mind that there are some important questions that must be answered. First, and foremost, is the method ethically sound? Second, will the questionnaire fully capture the context? Third, has it been tested and proven to generate reliable outputs; and fourth, what are the results going to be compared against?

Given these difficulties and risks it is best to use a method such as the Building Use Studies (BUS) survey, which is highly developed, well-proven, statistically sound, and which uses a database against which survey results can be compared. It also enables statistical data to be obtained along with anecdotes and comments. One-liners from occupants can often be just as insightful as the statistics. Phrases like 'It's a wonderful building to look at, but we keep having to go outside for air', can be devastatingly revealing.

With a high response rate – 100% is not unreasonable with planning, dedication and judiciously applied insistence on the part of the surveyor – a BUS survey can control for gender differences, responses from different areas of the building and extremes of opinion.

Once a survey of the existing building has been carried out, and the statistical and anecdotal results obtained, what is the next step? If the refurbished building has to inherit some of its existing constraints, then designers can do themselves (and the occupants) a huge favour by using occupant surveys to understand what people hate about the building, but also what they find virtuous about it, and design accordingly. It then becomes a matter of clearly communicating the design intentions, and any limitations of the retained elements – issues such as circulation routes that might be fixed.

Just because a building is old and ripe for refurbishment doesn't mean that it lacks virtue. Designers may be surprised when an occupant survey of what they consider to be a clapped-out building identifies aspects that the occupants appreciate and score highly. The usual qualities are an old building's ease of use, the legibility of controls and familiarity with the way the building responds during the seasons. People will have learned to live with the building's limitations and will act accordingly to keep themselves comfortable and happy.

4.1 Unmanageably complex light switches can alienate and confuse occupants when a simple light switch would have been more appropriate

Designers are trained to believe that thermal comfort is usually the most important determinant of occupant satisfaction. While that is true, occupants' tolerance to discomfort – not just of temperature but of many other factors – is more about how much control they have over their environment; and where they do not have control, how quickly and effectively control is exercised by others. People tend to be happier when a perceived need is met quickly, or when they are able to intervene to improve things to their satisfaction. It is not a question of maintaining a specific condition or set-point – the usual aim of both building designers and facilities managers who like to command and control – but rather about whether people have the opportunity to intervene to change things if conditions, or their needs, alter.

Findings from PROBE, and many subsequent post-occupancy studies, have shed stark light on the problems caused by complex electronic systems. So-called intelligent controls are considered vital technologies for the precise control of heating, ventilation and air-conditioning systems. Complex controls, which more often than not come with bespoke communication algorithms, are being used on what used to be simple devices, such as lights, vents and sky-lights (see Figure 4.1). The unmanageable complexity that comes with these controls can stress occupants to the extent that they feel alienated and unable to manage their working environment in the way they once did. Resentment, rather than gratitude, will be the result.

Designers who try to impose exactitude by introducing complex controls will not necessarily improve occupant satisfaction and energy efficiency. What they might gain in certainty of system control, they may lose if the refurbishment makes the building complicated, confusing or perhaps highly resistant to its occupants' attempts to run things the way they want. If people have good controls, in most situations it is better to provide environmental conditions – temperature, humidity, ventilation rates – which are 'good enough' rather than exact.

Designers who are fixated on automation of one kind or another, and who ignore the role of occupants in that equation, will run serious risks with the building's performance. The refurbishment could actually take the building backwards rather than forwards. So, before opting for such expensive and intrusive technologies, designers should focus on improving what the occupants already have, what they easily understand and have learned to operate, but which can nevertheless benefit from some improvement. This may simply be a case of providing better switches located in more appropriate locations rather than ripping it all out and imposing complex automation.

Designers are naturally wary of responding too directly to the findings generated by occupant surveys. Their reticence is understandable even if it is not wholly justified. The statistics and anecdotes from occupant surveys do not sit comfortably with the science of carefully constructed computer design models. The last thing a designer wants to do is alter a design strategy based on occupant feedback when survey results – or the way in which the survey was conducted – might be subject to error.

But this is to miss the point set out at the very start of this essay: building performance is not about achieving an absolute value, whether related to a benchmark or to a modelled prediction. Building performance is about what is realistic and desirable given the context. It is about understanding the purpose to which the refurbished building is to be put, identifying the needs and expectations of the end-users, and managing those expectations throughout the entire project. It does not only concern energy efficiency and lower carbon dioxide emissions, it is about building up an overall picture of what will be usable, manageable and maintainable, and applying that knowledge to every aspect of design and procurement.

THE IMPORTANCE OF REALITY CHECKING

Building performance evaluation is not a snapshot like POE, it is a longitudinal process that has to run from the inception stage of the project through construction, handover and occupation. This implies, of course, that the design and construction team will regularly revisit the ambitions and expectations expressed in the client requirements and the design brief and reality check them as the project evolves. As ideas turn into concepts and concepts into systems and systems into products, the project team needs to ensure that the objectives are not being compromised.

In the salami-slicing of responsibilities that is a feature of modern design and build procurement, the baton of responsibility can all too easily be dropped or hidden rather than passed on. This is why regular revisits to the project brief will ensure that any inconsistencies are noticed and addressed. Targets may have to change and expectations may have to be downgraded if other constraints – tighter budgets, for example – force the design team to accept systems that are not quite what they intended. In that case, the consequences have to be faced rather than ignored in the hope that the world will self-correct. It won't. It is far better that the performance metrics – however they are expressed – are downgraded in the client's requirements rather than becoming gradually and imperceptibly unrealistic, and a matter of surprise and resentment among the end-users when they inherit a building that fails to deliver what they were expecting.

The reality checking of design decisions should be systematic and regular. It needs to be integral with other project management functions, not an expendable add-on (see Figure 4.2 overleaf). The client should require it, the project manager should champion it, and the lead designer should lead it.

Generally, benefits can be categorised as short-, medium- and long-term:

- short-term benefits – immediate design decisions and facility maintenance and management issues resolved
- medium-term benefits – lessons learned fed forward to future projects for all teams involved (i.e. architect, construction team, consultants, building owners, etc.) closing the feedback loop
- long-term benefits – improving the long-term performance of buildings and justifying financial investment and major expenditures through monitoring and verification.

In the pressure cooker of a construction project, designers and constructors will not have time to reality check everything. At best, four or five critical items will be given full attention. But it is better to reality check a few things well, than try to reality check many inadequately. In any case, once the project team has got the hang of it, the halo effect may apply, and other aspects of the project might be considered more deeply without having to run all design decisions through a structured reality-checking exercise.

Unlike the well-defined building performance evaluation tools, such as occupant surveys and energy analysis reality checking is more of an idea than a prescription. The research and consultancy association BSRIA has defined one way of conducting a reality-checking process in its guidance document *BG27/2011 Pitstopping*, which, as its title implies, involves pulling in critical design decisions for a regular check against project expectations before injecting them back into mainstream procurement and construction.

The outputs from reality checking must be recorded and clear actions given to members of the project team. Some early adopters of Pitstopping are using

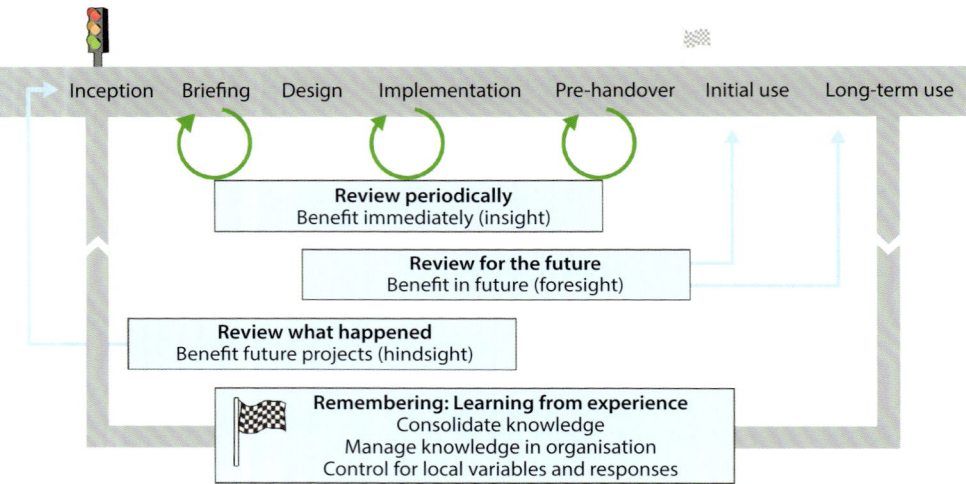

4.2 Reality checking and the importance of the feedback loop (after Bill Bordass).
The building performance evaluation feedback cycle, with reality-checking reviews

the project's operational risk register as a means of capturing the outputs. The risk register can be populated with the expectations and targets set at the project's early stages, and any deviations, changes and emerging knowledge compared to these benchmarks as procurement decisions are made. For example, it may be that once an idea becomes an actual product the designers realise that what they thought was fit-and-forget will be rather more fit-and-manage. In this case they have a choice: they can either ignore the operational complexities of the system and hope for the best, or give extra attention to the product's specification, installation and commissioning and plan for increased end-user training and familiarisation, with more end-user support before and after handover. The knowledge might usefully inform the requirements of the maintenance contract.

All of which leads logically to the final revelation in building performance evaluation: designers and other key members of the project team have to assume greater responsibility for building performance outcomes, and remain engaged until well after handover and beyond the traditional 12-month defects period. Whether this should fall within the professional's standard offering or be additional paid duties is a moot point. Either way, it is certainly a professional obligation that must be discharged if sustainable design is going to mean anything, and if the performance gap between design ambition and operational outcome is going to stand any chance of being understood and closed.

TAKING CUSTODY OF BUILDING PERFORMANCE

Emerging understanding of the gap between what designers intend and what actually happens begs a number of questions for the future of retrofit, and for building design generally. Whichever way one looks at it, the days when professionals could walk away after practical completion, and move on to the next job without much of a backward glance, are over. Project teams – and professional designers particularly – need to stay engaged to assist, troubleshoot and fine-tune their buildings. They should also embrace building performance evaluation to check the building's performance and use the results to help the building managers to improve performance.

The same tools of building performance evaluation used to inform the design can also be used to check the building's performance. However, there must not be a rush to judgement. Such is their importance in winning new work, that new and refurbished buildings are lauded almost from the moment they are completed. Before the days of the performance gap, this involved little risk. These days it is a dangerous business. Buildings rarely perform anywhere near their design targets from day one of occupation. They usually need 12 months to settle down, and often much longer before they can perform at their optimum, and even that is dependent on a lot of fine-tuning and seasonal adjustments.

Fortunately, there is now a process for this that construction teams can follow: Soft Landings. Described in the Soft Landings Framework, Soft Landings is an open-source, five-stage process, running from project inception through to extended aftercare.[4] It is designed to help project teams engage with performance outcomes. With Soft Landings, key members of the project team are required to provide professional aftercare for up to three years after handover. By remaining engaged with the building and the building operators, the aftercare team can support the occupants in their new home, spot emerging problems and help to resolve them before they fossilise into chronic shortcomings.

Soft Landings also provides a structure for periodic performance analysis of the building at set points during that three-year period, where the use of building performance evaluation can compare the building's emerging performance with the design intentions and targets. The results can inform the team where things need to be adjusted or recommissioned and other interventions made to improve the building's performance.

Table 4.1 (overleaf) shows how three key elements of building performance evaluation – operational energy consumption, human factors and physical measurements – map onto the 2013 edition of the RIBA Plan of Work and the Soft Landings graduated handover process.[5] Bear in mind that the steps and activities described in the table are not prescriptions. What you do, when you do it, and to what depth all depend very much on the context. As said earlier, it is best to take a graduated approach to analysis. Don't throw the kitchen sink at the building. Measure what matters and only get into the detail when you can't easily get to the root cause of an issue. The more you monitor, and the more things you monitor, the more data you amass and the more time you have to spend analysing it. That is ideal if you are an academic researcher for whom data is gold-dust but, in the real world of construction, time and money are limited, and the one thing you don't need is data-smog.

So what proof is there that everything described in this essay will deliver a better refurbishment? The fact is, of course, that although evidence is out there, not much is available in the public domain. Design teams like to play things close to their chest. Clients, too, are wary of publicity. But here is a shining example of how it can work: the Elizabeth II Court in Winchester, an office refurbishment for Hampshire County Council (HCC), shows what is possible given enough time, sufficient budget and a design brief that is informed by analysis and feedback from the existing building (see Case Study 2 later in this book). Ambitions for energy use and occupant satisfaction were well-informed and therefore realistic. The project serves as a useful template for other refurbishment projects, particularly as the key to reducing carbon emissions from the built environment lies with retrofit rather than with new build.

Table 4.1 Workflow table for retrofit

			Building Performance Evaluation (BPE) activities for retrofit projects		
RIBA 2013 Plan of Work	Soft Landings stages	Reality checking	Operational energy and carbon	Human factors	Physical measurements
0. Strategic definition					Identify the regulatory standards that will apply, and decide whether to go beyond them
1. Preparation and brief	Stage 1: Briefing. Identify all actions needed to support the procurement		Energy survey of the existing building. Set notional objectives for energy and environmental performance, record in client requirements	Occupant survey of existing building, workshops and structured interviews and design quality assessments	Identify targets for physical performance (typically fabric quality, insulation levels, level of airtightness, and permeability of specific elements). Select calculation or laboratory values
2. Concept design	Stage 2: Design development, to support the design as it evolves		Build energy model with notional regulated and unregulated loads, hours of operation. Revisit and inform client requirements	Present findings of occupant surveys and assessments to occupants and stakeholders. Record end-user expectations	Ensure theoretical simulation and design models meet the design targets set in the client's requirements, and that the data are easily accessible for comparison with as-built
3. Developed design		Reality check 1: The entire scheme design	Revisit energy model with loads as they become available and capacities of favoured HVAC systems. Check against design brief ambitions. Identify uncertainties and disparities. Refine energy targets in the light of any changes to client's requirements and developed design	Use end-user expectations to set human factors targets, e.g.: space densities, speed of response times by facilities managers, controls' usability requirements	Check client's requirements against design; check for buildability issues and revise targets where necessary (for example U-values, emissivity, airtightness, specific fan power, cavity widths, bulk density of insulation)
4. Technical design		Reality check 2: Technical reality check. Focus on critical systems	Revisit energy model as specific systems become known in the technical design phase; refine hours of operation; add plug-in power assumptions. Check against client requirements and design brief and take account of any changes and differences	Refer to human factors requirements to inform operational risk register items. Review all control interfaces and check for usability and manageability against expectations and abilities of relevant end-users	Plan for intermediate sampling and testing of elements against the sequencing of works, and the timing of periodic tests (for example, the vapour/air barrier). Decide on acceptable range of variation and error for in-situ measurements. Include in contractors' scope of works
5. Construction	Stage 3: Pre-handover. Prepare for building readiness. Provide technical assistance	Reality check 3: Tender award stage reality checking of key items	Revisit and refine energy model with known capacities of actual products selected by contractors. Refine operational profile and plug-in power assumptions	Manage occupiers' expectations as the project progresses, with feedback sessions with stakeholders and occupiers' representatives. Involve specialist occupiers as they become known (for example FM and catering contractors)	Perform intermediate sampling and testing as agreed at Technical Design. Decide on appropriate response if results are out of range, consider effect of remedial interventions on project programme
6. Handover and close out		Reality check 4: Pre-handover reality check of key items	Check energy model against installed systems, coordinated with the commissioning process. Check occupier's intended hours of operation and plug-in loads and update model	Ensure staff training and familiarisation is carried out and repeated as necessary. Review all operational guidance	Perform intermediate sampling and testing as agreed at Technical Design; ensure tests are coordinated (and consistent) with the commissioning plan
7. In-use	Stage 4: Aftercare in the initial period. Support in the first few weeks of operation	Reality-check 5: Post-handover sign off review	Adopt voluntary Display Energy Certification and/or Landlord's Energy Statement as appropriate. Update energy model at month 12 and report initial energy performance and carbon emissions	Support occupiers in initial 6–8 week period of aftercare post-PC. Regular walkabouts by design professional to engage with occupants, explain system operation; spot any emerging problems, report and help resolve	Assess as-built drawings. Perform in-situ tests within defects warranty period (local U-value tests, thermography survey, local checks for air leakage)
	Stage 5: Years 1 to 3. Aftercare, monitoring, review, fine-tuning, and feedback		Conduct formal energy survey in first post-occupancy evaluation no sooner than month 12 post-completion (later if phased occupation or fit-out works). Choose relevant energy benchmark for comparison purposes. Use findings to inform systems improvements and adjustments where necessary. Report second year operational energy consumption in POE within months 24-36	Perform occupant satisfaction surveys (permanent users) no sooner than 12 months post PC. Adopt separate survey and/or structured interviews for transient users where appropriate. Respond to findings, repeat survey for second POE within months 24–36	Consider deeper physical monitoring where post-occupancy evaluation finds problems that cannot be immediately explained (typically using thermography, air leakage tests, tracer gas analysis, temperature, humidity and indoor air quality measurements)
Documents and weblinks	*BSRIA BG4/2009 The Soft Landings Framework.* Other guidance from www.softlandings.org.uk	*BSRIA BG27/2011 Pitstopping – BSRIA's reality-checking process for Soft Landings projects*	CIBSE Technical Memorandums *TM22, T46* and *TM54* provide energy assessment tools and guidance on benchmarks. Also see www.carbonbuzz.org	Design Quality Indicators available from CIC. Building Use Studies (BUS survey) details from www.busmethodology.org.uk	Guidance on physical testing is available from many sources, including RIBA, BSRIA, CIBSE, CIC, and others

The building comprised North, West and East blocks on a podium car park. HCC decided to refurbish the entire site, excluding parking, and turn it into an exemplar of sustainable and energy-efficient office space. Sustainability and energy efficiency were key project objectives for the refurbishment, which set out to meet, and in some areas exceed, the requirements of Part L of the 2006 Building Regulations.

The three-storey East block, built in the 1960s, is typical of local authority offices of the period: heavyweight concrete construction with prefabricated concrete panels, and single-pane glazing with horizontally pivoting openable windows for natural ventilation. The East block was studied under the Carbon Trust's Low Carbon Buildings Accelerator (LCBA) research programme. The condition and thermal performance of the fabric was poor by contemporary standards: overall facade U-values varied between 1.5 and 5 W/m²K. The building was also not particularly airtight. Lighting was by fluorescent luminaires fitted in a suspended ceiling. This not only hid the building services, but also disconnected the building's thermal mass from the occupied space. The lighting control zones were too large for any energy-saving measures to be effective. In 2006, a BUS occupant satisfaction survey of the East block showed that the occupants were very unhappy.

The form of the building, with long, narrow floorplates and manageable floor-to-floor heights, lent itself to a low energy refurbishment. The design team wanted to establish an environmental strategy that was sufficiently robust and flexible to accommodate change with an emphasis on sound environmental engineering rather than over-reliance on sustainable technologies. The building services were required to be simple, without overly complex control systems.

Early energy calculations suggested 59 kWh/m² per annum for fossil fuel, and 34 kWh/m² per annum for electricity. The targets were subsequently refined to 57 kWh/m² per annum for fossil fuel and 66 kWh/m² per annum for electricity (total 121 kWh/m² per annum). The combined carbon dioxide emission target of 39 kgCO₂/m² per annum was equivalent to a 10% improvement over the *Energy Consumption Guide 19* (*ECON 19*) good practice performance for a

hybrid office, with a 15% Type 3 component. The carbon dioxide emission target was subsequently reduced to 35 kgCO₂/m² per annum as more information became available during design development.

When these targets are compared with the estimated figures for the old building – between 320 and 335 kWh/m² per annum for gas and electricity, and combined carbon emissions of 90–92 kgCO₂/m² per annum – the scale of the challenge set by HCC becomes clear.

All the buildings on the site were extensively remodelled. The top floors of the West and East blocks were cut back to reduce both the overall massing and the perceived height of the building. The car park space at podium level under the East block was changed into office accommodation, with the similar area under the West and North blocks used for a new IT suite and ancillary spaces. This included a new reception, a café and auditorium pavilions. These changes saw an increase in the accommodation from around 9,000 m² to nearly 14,000 m², of which the East block represented about 25%.

The floor-to-floor heights on the office floors are approximately 3.35 m, as determined by the dimensions of the original structure. The removal of the suspended ceiling exposed the original concrete slab of 880 mm coffers, increased the effective floor-to-ceiling height and released the thermal capacity of the building to assist in moderating internal temperatures. Cables and air supplies were run within a 320 mm raised floor.

A hybrid, mixed-mode approach was adopted and the design developed so that fan energy consumption could be minimised. On the lower floors, the street elevation is equipped with ventilation chimneys, located at regular intervals along the facade. Each extract chimney serves one of the three lower floors. The motorised windows and the chimney sashes are automatically controlled to vary the amount of natural ventilation. Openable windows were provided to give occupants the freedom to control their ventilation needs, enabling them to trade off air movement, temperature, air quality and outside noise as they see fit.

The East block of Elizabeth II Court was completed in December 2008. HCC and the Carbon Trust team carried out performance monitoring of this block over a 12-month period, from November 2009 to October 2010. Although too early to be considered a true Soft Landings project, Elizabeth II Court nonetheless adopted many elements of Soft Landings aftercare. The programme included a three-month initial period of 'light touch' monitoring, starting in November 2009. This period was eventually extended to May 2010, followed by two months of detailed monitoring in June and July 2010. A final three months of light touch monitoring ran up to October 2010.

As the refurbished East block has mixed-mode ventilation and some mechanical cooling, no direct equivalent energy or carbon performance benchmark is available. At carbon factors of 0.194 for gas and 0.422 for electricity (to maintain consistency with Part L) the target carbon dioxide emissions were 35 $kgCO_2/m^2$ per annum. The 12-month monitored results came in at 131 kWh/m^2 per annum; a mere 7% higher than the original design target of 121 kWh/m^2.

With initial gas consumption at 54 kWh/m^2 per annum, the East block of Elizabeth II Court initially performed slightly better than the original design target. At 77 kWh/m^2 per annum, the total electrical consumption of the East block of Elizabeth II Court was 17% higher than the original design target, which the building operator believed could be improved upon over time. At 43 $kgCO_2/m^2$ per annum, the initial carbon dioxide emissions were 10% higher at the outset than the target, but this can be considered admirably close for the first 12 months of operation.

Occupant satisfaction with the East block of Elizabeth II Court post refurbishment was measured by repeating the BUS occupant satisfaction survey in July 2010. It showed a huge improvement in the occupants' view of the refurbished building, which now scores particularly well in terms of design, occupant needs and image.

MOVING FORWARD

So what does the evidence from such building performance evaluation tell us? It shows that by adopting the principles described in this essay, project teams can deliver a better standard of refurbishment. Client and end-user expectations can be ascertained and managed by using freely available tools, and energy use can be predicted with a fair degree of accuracy (as long as everything is counted – not just regulated loads to achieve compliance with the Building Regulations, but plug-in loads and hours of operation as they become known). And, with Soft Landings, a mechanism now exists for project teams to stay engaged to follow though and fine-tune buildings, with the option of carrying out seasonal commissioning, not as a separate exercise when something goes wrong, but as part of the project team's standard professional services. We now have the means at our disposal to close the so-called 'performance gap', which has become a hot topic in the race to meet our zero carbon targets.

However, a word of caution is required. The term 'performance gap' largely defines the chasm between what design teams say is possible and what is subsequently delivered. Where a performance gap is found to exist – whether in energy consumption, carbon dioxide emissions or something else – the implication is that it is the fault of the designers. But the truth is that only a foolhardy designer makes firm predictions at a project's early stages. Instead, what they do is make calculations. They make broad-brush estimations. Whether this is done on the back of a cigarette packet or generated by a computer simulation tool doesn't really matter. Early estimations will contain huge assumptions about plant loads, operating hours, intensities of use and construction quality.

The problem with publicising estimations to the client is that they have a nasty habit of being read as predictions, and those predictions then have an equally nasty habit of becoming accepted targets. Once the design has been signed off, the estimations will come under stress from many influences. Value engineering, product substitution and unchecked changes in specifications will undermine design integrity. In the absence of continual risk assessment and sensitivity analysis, loads associated with ICT and servers will creep in unnoticed. A fixation on time, cost and programme at the expense of quality will compound the issues. If commissioning is poor, handover is rushed and operational training is non-existent, is it any wonder that the performance of the building will be vastly different from the design intention?

Clients need to become wise to the fact that performance is not solely the responsibility of the design team. The entire project team – the client, the designers, the main contractor, the project manager, the cost consultant and the specialist subcontractors – all must assume responsibility for building performance. Sustainable retrofit is not about technical innovation, low and zero carbon technologies and environmental rating schemes. The real revolution in sustainable retrofit needs to be in the development of a more intelligent procurement process based on shared risk and collaborative working. And, in that endeavour, we have just begun to climb a mountain.

ENDNOTES

1 B. Bordass, R. Cohen and J. Field, *Energy Performance of Non-Domestic Buildings: Closing the Credibility Gap* (2004) Building Performance Congress, Frankfurt, April 19-24, 2004. Available at www.usablebuildings.co.uk/Pages/Unprotected/EnPerfNDBuildings.pdf (accessed 1 November 2013).

2 B. Bordass, R. Cohen, M. Standeven and A. Leaman, Assessing Building Performance In Use 3: Energy Performance of the Probe Buildings, *Building Research and Information*, Vol. 29, No. 2 (2001), pp. 114–128.

3 Outputs from the Carbon Trust's LCBP and LCBA programme can be found at www.carbontrust.com/resources/guides/energy-efficiency/low-carbon-buildings-design-and-construction (accessed 25 November 2013). See also, Carbon Trust, *Closing the Gap: Lessons Learned on Realising the Potential of Low Carbon Building Design*, London: The Carbon Trust (2011).

4 R. Bunn, *How to Procure Soft Landings – Specifications and Supporting Guidance for Clients, Consultants and Contractors*, BSRIA BG 45/2013 (2013).

5 *RIBA Plan of Work 2013*, Royal Institute of British Architects (2013). Available at www.ribaplanofwork.com (accessed 25 November 2013).

chapter 5

RAJAT GUPTA & MATT GREGG

EVALUATING RETROFIT PERFORMANCE

A PROCESS MAP

INTRODUCTION

It is common for claims about high performance to be made early in projects for both new and retrofitted buildings. However, it is also common for the actual performance of buildings to fall far short of what was imagined in these initial stages. Fortunately, rigorous methodologies exist to measure building performance and they can be adapted specifically to retrofit projects.

FACING PAGE Foundry Studios, North London. See Case Study 3

By 2050, in the UK, a majority of the 26 million homes and about 1.8 million non-domestic buildings that were built before any concerns about CO_2 emissions were raised will still be standing. They are currently responsible for 26% and 18% of UK CO_2 emissions respectively.[1] Furthermore, although the non-domestic stock is more modern than the housing stock, half of the commercial and industrial properties in 2005 were built before 1940 and only 9% after 1990.[2] Despite the far greater impact of existing buildings, the primary focus of energy efficiency and carbon reduction efforts has been on new buildings. As this imbalance is corrected, and the retrofit programme gathers pace, a systematic and empirical approach to providing an evidence base for the performance of buildings in use becomes essential. Building Performance Evaluation (BPE) which involves feedback and evaluation reviews at every phase of the building delivery, from strategic planning to occupancy, offers a robust approach for establishing which retrofit interventions are effective and which interventions fail to deliver the targets set, and why they do so.

Retrofitted buildings are as prone as new buildings to claims for 'high performance' or 'sustainability' backed with little evidence. The impact of this tendency is exacerbated by the 'performance gap' – the phenomenon of the actual energy use, and consequent greenhouse gas (GHG) emissions, of buildings being typically double the design intent. The Government programme the 'Green Deal for Business' sets out to help prepare non-domestic landlords for the 2018 deadline, after which it will be illegal to rent out a property that has a rating lower than an 'E' on its Energy Performance Certificate (EPC). However, the performance gap threatens to make such measures virtually ineffective. Without BPE cyclically feeding in lessons learned at the design, construction and operation stages, the performance gap between design intent and actual performance could become ubiquitous as Green Deal type programmes become a major driver to enable low energy retrofitting.

EVOLUTION OF BUILDING PERFORMANCE EVALUATION IN THE UK

The need to evaluate the in-use performance of low energy buildings has led to the evolution of Post Occupancy Evaluation (POE), a process that historically takes place after the building's completion as a diagnostic tool of the building's actual performance. In the late 1960s, POE began as one-off case study evaluations, focusing on the residential environment of urban renewal projects that took place in North America and Western Europe, to investigate and resolve unexpected architectural and social problems, such as sick building syndrome.[3] In the years to follow, POE studies were used to evaluate a variety of building types, including educational and health institutions, prisons, offices and other commercial buildings.[4]

Recognising the importance of performance feedback throughout the whole life-cycle of a building, Preiser and Schramm introduced the term 'building performance evaluation' (BPE) to address feedback loops in different stages of a building's life cycle.[5] BPE involves feedback and evaluation reviews at every phase of the building delivery, from strategic planning to occupancy, adoptive reuse and recycling.[6] Eventually, as the concern over performance in the building sector grew to encompass energy and climate issues, BPE and POE became internationally accepted methods for discovering and providing solutions for performance issues and inefficiencies in buildings.[7–11]

In the UK, beginning in 1995, a Government-funded research project called PROBE (Post-occupancy Review Of Buildings and their Engineering) originally sought to provide feedback to building services engineers on success, difficulty and disappointment in light of the growing importance of energy efficiency in design, a relatively new facilities management (FM) profession, new interest in indoor environmental health, and technological improvements in computer modelling, building fabric and glazing systems and more efficient air-conditioning, etc. The approach combined energy assessment, occupancy evaluation, questionnaires and airtightness testing. The findings revealed the methodology to have a much wider value in informing improvement in design, commissioning and brief preparation.[12]

Technically, POE studies (a subset of BPE) involve systematic collection and evaluation of information about the performance of a building in use. Data collected can include measured information, such as energy consumption, temperature variation, lighting levels, acoustic performance, etc., and survey data from the perspective of the occupants regarding issues such as comfort, aesthetics, occupant satisfaction and management. POE is about getting useful results quickly and in a robust and inexpensive manner in order to solve problems and provide feedback. According to the Usable Buildings Trust (UBT), long-term comprehensive monitoring is seldom necessary other than for research purposes.[13] Given the high likelihood that occupant behaviour will have an effect on energy efficiency in most buildings, it is important to include evaluation of user comfort and satisfaction within POE studies. Without this social aspect, performance gaps between expectations and outcomes cannot be significantly reduced.

In 2004, the methodology of Soft Landings was developed from a couple of concepts including 'sea trials'[14] with input from UBT's FROBE findings with the intention of easing the transition from completion to in-use stage, with particular focus on the widespread problems revealed through well-documented experience of POE in the non-domestic sector (the PROBE studies). The Soft Landings approach, an evolution from POE to a closely managed, long-term interventionist form of BPE, includes design review, commissioning and handover observation and comprehensive aftercare and is applicable for new construction, retrofits and simple alterations,[15] thereby embedding performance evaluation within the entire life-cycle of buildings.

Over the past decade, BREEAM (Building Research Establishment Environmental Assessment Method) has also been developing a performance evaluation tool for buildings in use. BREEAM In-Use is a certification programme designed for the purpose of managing and reducing the running costs and environmental impact of existing non-domestic buildings through a holistic approach covering (though not limited to) energy, water, materials, transport issues and land use and ecology. Independent third party confirmation allows

the changes or 'claims' to be verified and rated on a scale that takes account of all the environmental variables included in the assessment.[16]

Currently, the best resourced force in performance evaluation in the UK is the UK Government's innovation agency, the Technology Strategy Board (TSB), which has been funding various BPE programmes since 2008. The TSB's BPE approach demands a combination of quantitative and qualitative data, together with expert analysis and insight. Throughout the evaluation period data is not just captured, but interrogated in a forensic fashion, informing additional data collection and lines of enquiry. Participants consider how performance can be improved and where possible, implement changes (especially low-cost ones affecting operation, control, management and user awareness) and measure their effect. In all cases, projects include a comparison of the designed (i.e. predicted) performance with delivered (actual) performance. The TSB's BPE approach focuses on two distinct periods of investigation:[17]

1. 'Post construction and early occupation' – 'as built' performance of the building envelope and installed equipment, effectiveness of the handover process and the occupants' initial reactions.
2. 'In-use and post occupancy' – performance of the building over an extended period of time, after fabric and systems have stabilised and the occupants are more familiar with the building.

Retrofit BPE projects should ideally be subject to a pre-retrofit evaluation mimicking both segments 1 and 2 on a shorter scale. This will help to establish the baseline performance of the building from both a technical and an occupant's perspective. Savings from retrofitting interventions can also be measured against this baseline performance.

In 2010, the TSB, jointly with the Department for Energy and Climate Change (DECC), developed a competition for 'Energy-efficient Whitehall' to begin testing the opportunities for retrofitting non-domestic buildings in anticipation of the implementation of the Green Deal. The competition was created to explore solutions to improve energy efficiency, through innovative retrofit solutions that reduce both the demand for energy and CO_2 emissions, in existing government office buildings in Whitehall, London. The scope of the retrofit work included the potential for in-use energy reduction methods (e.g. improving the performance, control and management of mechanical and electrical systems), supporting behavioural change by management and users and fabric modifications. The project proposals were required to include estimates of the energy savings and to evaluate the savings through post-occupancy assessment and monitoring for one year from 2011 to 2012.[18]

After the exploratory Energy-efficient Whitehall competition, the TSB and DECC funded a non-domestic retrofit BPE competition in 2012 called 'Invest in Innovative Refurbishment'. To align with the focus of the Green Deal in the non-domestic sector, contracts were awarded based on best energy efficiency expectation and best potential for large-scale roll-out of the methods used. The scope of potential retrofitting approaches (with an emphasis on innovation) remained similar to the Whitehall competition and includes a 12-month mandatory monitoring and POE stage, which will conclude in 2014 (Round 1 projects) and 2015 (Round 2 projects).[19]

BPE is invaluable for identifying, quantifying and resolving the gap between 'as designed' and 'in-use' performance through a systematic collection and analysis of qualitative and quantitative information related to fabric performance, energy performance (which is influenced by a range of factors from equipment to occupancy) and environmental conditions. BPE involves feedback and evaluation reviews at every phase of the building delivery from strategic planning to occupancy, adaptive reuse and recycling.[20] In addition, the Soft Landings framework provides extensive aftercare assistance at occupancy stages, identifying and actively closing the gap found there.[21] When retrofitting buildings, BPE ideally includes a pre- or inter-planning stage in which feedback and evaluation are performed to establish pre-existing patterns in occupancy and management.

BENEFITS OF BPE FOR RETROFITS: ADDRESSING THE LIMITS AND BARRIERS

The main purpose of BPE is to maximise the intended efficiencies (minimising unintended issues) by closing the performance gap and contributing to future closure through the learning process (minimising future unintended consequences). BPE has short-, medium- and long-term benefits. Specifically for retrofitting existing non-domestic buildings, BPE offers a range of advantages as detailed below, for designers, clients and users, which can lead to reduction in energy usage, utility bills and CO_2 emissions:

- pre-retrofit briefing, which can aid in identifying ideal strategies for performance improvement
- contextualised performance outcomes – detailed hypotheses based on experience that can be applied in future buildings
- better communication, transparency and accountability between design, supply, construction and commissioning
- improved handover, aftercare and fine-tuning
- greater attention to controls and details which affect efficiency and user interaction
- quantifying (through monitoring and verification) and delivering real returns on investment/reducing risk (e.g. improvements in environmental conditions that have occurred as a result of retrofitting)
- ascertaining occupant satisfaction with the change (retrofit).

Despite these benefits, there are also considerable barriers and challenges to the implementation and success of BPE which must be overcome:[22, 23]

- BPE and Soft Landings are not included in conventional financial plans for development
- depending on objectives, BPE can require a considerable amount of time
- professionals often do not like to have their work judged or its limitations exposed
- resistance to change in practice

- timely and organised dissemination can be difficult
- dependence on data availability – sometimes the required data (e.g. fuel bills or past energy use) is not retained and is unavailable
- occupants can be unresponsive to questionnaires.

To address these barriers and undertake an effective EPE study, the following measures are recommended during set-up, execution and dissemination of BPE:

- Develop a clear statement about what is to be achieved by conducting and applying a BPE. The links between evaluations and stated requirements have to be explicit and easy to trace.
- Match data collection and analysis to the available time, budget and resources.
- Determine appropriate methodology (presented in the following section). BPE methods for domestic, non-domestic, new-build and retrofit can vary.
- Inform respondents and/or occupants about the purpose and importance of their involvement and how the data will be used.
- Architects and engineers can assist in the integration of the BPE process into the design of the building by:
  o planning sub-metering into the systems design
  o working with the contractor to build mock-ups of difficult or unusual details to ensure performance and documenting results
  o verifying at an early stage that the commissioning agents understand what is expected and what will need commissioning.
- Interpret and synthesise BPE findings in useful ways.
- Make results widely available for action by different stakeholders (e.g. written reports, videos and websites).
- Use BPE findings to facilitate organisational learning in order to improve building performance under routine conditions and respond to changing conditions quickly and effectively.
- Develop an effective method to disseminate the consolidated data – a record of past projects, experience and results may serve as a justification for future BPEs.

BPE PROCESS MAP FOR NON-DOMESTIC RETROFITS

To enable a wider uptake of BPE studies, a robust and comprehensive framework or process map is proposed here for undertaking and evaluating non-domestic retrofits. The process should start at the *pre-retrofit stage*, continue during *retrofit* and *post-retrofit* and conclude at the *in-use stage*. Table 5.1 provides a workflow for the BPE retrofit, while Table 5.2 lists the tools and methodologies that are commonly adopted for undertaking BPE studies at the respective stages of retrofit. Each table is set out in order of implementation, reading down the page.

Table 5.1 Workflow of non-domestic retrofit BPE

	Existing occupancy and management evaluation – POE	As-built drawings and specifications review, energy analysis (fuel bills, sub-meter readings), spot measurements, air-permeability test, BMS details, interviews and walkthroughs with facilities manager (FM), occupants, Building Use Survey (BUS) questionnaires
Design and construction		
	Design and construction audit	Review of drawings, SBEM (Simplified Building Energy Model) calculations, environmental assessment standards, interviews/feedback from design and construction teams to compare design intentions to built reality (and later performance)
	Fabric performance testing	Heat loss test, air-permeability test/smoke-based air leakage test, infrared thermography, in-situ U-value measurements
	Systems installation and commissioning review	Installation and commissioning checks, measurement of performance and energy use of environmental control systems (e.g. mechanical ventilation, etc.)
	Control interface(s) review	Review of the usability of control interfaces
	Handover and written guidance evaluation and review	Review of handover process and user guide documentation
	In-use evaluation/POE	Review of occupant and manager feedback – walkthroughs, interviews and BUS questionnaires
		Years 1–2 monitoring and occupant feedback review
Final analysis, feedback and dissemination		

Table 5.2 Common retrofit BPE tools and actions[24–26]

Tool/action	Stage	Measures	Reveals/provides
Stakeholder induction	Project initiation	Involvement	Roles and procedures
Review of analogues	Project initiation	Past experience	Lessons learned and key findings of similar BPEs
Collection of past energy data (e.g. bills)	Pre-retrofit BPE (ideally before design stage)	Pre-retrofit energy use patterns	Baseline for post-retrofit comparison
Evaluation of as-built drawings and other documentation (including performance target)	Pre-retrofit BPE (ideally before design stage)	Existing conditions	An understanding of the pre-retrofit design, intentions and targets
Facilities management questionnaire/interview	Pre-retrofit BPE (ideally before design stage) Initial occupation (post-retrofit)/in use	Facilities manager's operational habits, patterns and manageability	Management issues and concerns – analysed against energy use
Walkthrough energy survey (photographic or video documentation). Use bespoke checklists	Pre-retrofit BPE (ideally before design stage) Initial occupation (post-retrofit)/in use	Equipment and services performance27	Equipment and services type, rating, operation, etc.
Metering/sub-metering (and/or BMS data collection) (e.g. boiler, ventilation system, renewable technologies) CIBSE *TM39: Building Energy Metering* Also water consumption metering	Pre-retrofit BPE (ideally before design stage) In-use (post-retrofit), ongoing	Energy use of individual zones/equipment/circuits/appliances (individual power meters can be used for appliances)	Baseline for post-retrofit comparison. Sub-metered (disaggregated) energy use: isolation of performance and problem solving
CIBSE *TM22: Energy Assessment Tool* Weather normalisation (degree days)	Pre-retrofit BPE (ideally before design stage) In-use (post-retrofit), annually	Energy use disaggregation	Energy profile and energy use breakdown: standard tool for benchmarking and reporting
Environmental spot measurements, ongoing on-site monitoring and climatic data collection (weather station)	Pre-retrofit BPE (ideally before design stage) In-use (post-retrofit) spot measurements – seasonally or annually/monitoring ongoing	Internal: temperature, relative humidity (RH), CO_2 (proxy for indoor air quality (IAQ)), daylight factor, and noise levels. Advanced IAQ: ventilation flow rate, volatile organic compounds (VOCs), NO_x, CO, etc. External: temperature, RH, and solar irradiance	Pre-/post-retrofit comparisons, comparisons with occupant perception and isolation of irregularities with implications for energy use
Occupation monitoring	Pre-retrofit BPE (ideally before design stage) In-use (post-retrofit) monitoring ongoing	Use of doors, windows, other openings and occupancy levels by passive infrared (PIR) detection	Occupation patterns and habits – to be analysed against energy use, IAQ, opinion, etc.
BUS questionnaire, interviews and walkthroughs for occupants	Pre-retrofit BPE (ideally before design stage) In-use (post-retrofit) seasonally or annually	Occupant satisfaction, habits, and concerns	Opinion on aesthetic, comfort, noise, air quality, perception of health and control, etc. Pinpoint issues, problem resolution

Tool/action	Stage	Measures	Reveals/provides
Occupant participation tools, e.g., journaling, photographic and video audits (by occupant) and focus groups	Pre-retrofit BPE (ideally before design stage) In-use (post-retrofit) seasonally or annually	Occupant satisfaction, habits and concerns	Occupant habits, opinion and interaction. Pinpoint issues, problem resolution
Infrared thermography (effective in combination with air permeability test and assists in locating ideal locations for smoke tests and U-value measurements)	Pre-retrofit BPE (ideally before design stage) Post-completion (fabric testing) BPE final analysis	Qualitative visualisation of surface temperatures	Heat loss, thermal bridging, gaps in insulation, changes in insulation, areas of in/exfiltration, etc. Identify areas in need of improvement or repair
Air permeability test (mean of pressurised and depressurised values) Air leakage identification (e.g. smoke pencil)	Pre-retrofit BPE (ideally before design stage) Post-completion (fabric testing) BPE final analysis	$m^3/(h.m^2)$ at 50 Pa: air leakage per square metre of building envelope, including all wall, roof and floor areas that are exposed to the external environment	Provides air-permeability rating. Smoke pencils can be used to find paths of infiltration and identify specific weak points in the building envelope
Co-heating test (small scale non-domestic only)	Post-completion (fabric testing) – strictly unoccupied	Measures the daily heat input against the daily difference in internal and external temperature resulting in the heat loss coefficient	Heat loss from both fabric and uncontrolled ventilation
U-value test (heat flux sensors)	Pre-retrofit BPE (ideally before design stage) Post-completion (fabric testing)	U-values of wall assemblies	Helps identify the actual value of insulation improvements. Can be used for comparing post-retrofit results to pre-retrofit and 'as-designed' information
Guide for reviewing controls for end users[27]	Post-completion	Ranking of controls (e.g. clarity of purpose, intuitiveness, labelling, ease of use, etc.)	Appropriateness of design and implementation of controls. Can be compared with occupant opinion
Handover evaluation	Post-completion	Review of process and all documentation (e.g. occupant user guide, FM guide, building log book, etc.)	Areas in need of improvement and where further education required for managers, FM, occupants, etc.

PRE-RETROFIT STAGE

Existing occupancy and management evaluation

Understanding how the building is managed can assist in performance predictions rather than making assumptions about an idealised management scenario. Pre-retrofit BPE can go far beyond simply establishing management practices and can be used to inform the design through the evaluation of user practices, existing control interface and system issues, and occupant expectations. Furthermore, establishing existing energy use, air-permeability and environmental performance measurements (baseline performance) is useful for benchmarking the improvement, savings and benefits post-retrofit.

The pre-retrofit stage is where BPE of a retrofit can be most clearly distinguished from a new-build BPE:

- For the client, additional or completely different needs can be revealed for building redesign from a pre-retrofit analysis of the building, thereby improving the brief. This can include thermal comfort issues that were previously unknown, etc.
- For the designer, this stage can clarify design priorities and ultimately improve the performance outcome.
- For the BPE evaluator, pre-retrofit data provides valuable benchmarking information. It also clarifies focus points for the post-retrofit analysis, giving clear direction to the analysis and purpose of the study.
- As the extent of retrofit can vary, the results of the pre-retrofit BPE can inform either improvements concurrent with occupation or a full overhaul, or deep retrofit of the building.

DURING RETROFIT

Design and construction audit (during the retrofitting process)

The design and construction audit is the comparison of the initial design intentions against the constructed reality to discover the rationale behind changes and to evaluate impact.[28] This stage includes:[29]

- design drawing, documentation and specification review (including metering/sub-metering strategy); particular attention should be given to where and why changes were made during construction
- review of the arrangements for managing delivery of design intent
- site visits, including photographic documentation during and after construction
- on-site build quality testing with documentation
- review of commissioning plans and procedures – this will assist when completing the systems installation and commissioning review
- modelling, calculation and simulation review; particularly SBEM (required for non-domestic building regulations compliance)
- capturing findings and identifying possibilities for improvement – discussing these with client, occupier, design and building teams
- construction completion walkthroughs and interviews with developer and designers.

Often, much is revealed regarding the lack of communication between trades, miscommunication of expectations, and incomplete planning and integration of systems, all potentially contributing to a gap in performance.

POST-RETROFIT EARLY OCCUPANCY STAGE

Fabric performance testing

This stage, which takes place as key elements of construction are completed, can include the following:

- whole building heat loss test or co-heating test: typically for small (<300 m²) non-domestic buildings or spaces
- in-situ U-value measurements: heat flux measurement of representative fabric – for a duration of two weeks, with measurements at five-minute intervals
- air-permeability test/airtightness test: used to assess heat loss due to air moving in and out through the building fabric alone, best performed as soon as the building can be sealed for the test, to allow remedial action to be taken within programme
- infrared thermography: assists in locating construction failure, major areas of heat loss and air leakage by visualising thermal radiation from or into a building. Thermography is an indispensable aid to visualising the data collected and completing the story told by the heat loss, U-value and air-permeability tests.

Systems installation and commissioning review

A post-completion stage, during which the precision of installation and commissioning and the operational use of the systems are reviewed, provides an opportunity to correct faults. Faults in design, construction, installation, commissioning and coordination can also be documented and elucidated for future improvement. Involving facilities managers in this review can provide a deeper awareness of system integration and possible failures, and promote FM ownership of future failures and corrections.

Control interface(s) review

The control interface is the point where the occupants are faced with the building's technology or the BMS. The controls affect user satisfaction, comfort and perception of comfort, cost effectiveness in managing localised problems and energy efficiency. Problems often arise due to a disjunction between the design/installer side and the expectations of the end user. They can include:

- unintuitively located controls
- lack of labelling, confusing colours and direction of control
- not enough fine-tuning for comfort control
- insufficient fine gradation in zoning of controls
- no tangible feedback – users need to know that settings are being changed on the control as a result of their actions
- no obvious response – users need to know that the control is working.

Findings should be fed forward to suppliers/controls manufacturers.

Handover and written guidance evaluation and review

The handover stage is the pivotal point between the post-construction and occupation phases, essential for the building management staff to be inducted in the use of controls. Ideally, control demonstration and building user guidance should also be given to the occupants by the owner or FM. Written guidance should include a guide for occupants, a technical guide for FMs and the operations and maintenance manual. CIBSE's *TM31: Building Log Book Toolkit* can be followed to assess the completeness of the building log book given to the owner, providing summary information about the retrofitted building, its building services and their maintenance requirements. The Soft Landings framework provides extensive pre-handover preparation, and requires continuing access to aftercare representatives who will provide ongoing support for an extended period of time.

POST-RETROFIT IN-USE STAGE

In-use evaluation

Post-retrofit, the building is in a new stage of life, the energy analysis and benchmarking takes place after occupants have settled into the building and have become familiar with it. Typically, the in-use BPE can be divided into five essential elements:

1. Collection and analysis of essential background information and available data: drawings, specifications and predicted resource consumption.
2. Continuous monitoring/metering and sub-metering of energy (with results as disaggregated as possible), water and environmental (internal and external) variables (e.g. temperature and humidity) using data loggers or transmitting monitoring equipment in strategic areas.
3. Walkthrough evaluation of all elements from systems to controls and occupant behaviour.
4. Questionnaire surveys (e.g. BUS) and interviews of building users, to assess user satisfaction and behaviour. Beyond questions regarding thermal comfort, lighting, ventilation, air quality and noise there are questions which will determine the user's perception and satisfaction in general. For example, does the building meet the users' needs? Does it meet the business requirements? In addition, the following areas can be considered:
   o Building controls: assessing the occupants' indoor environmental satisfaction (covering thermal comfort, air quality, ventilation and daylight), their current levels of environmental awareness, and use of controls, comparing the results with monitored data and the post-completion control interface review.
   o Building design and aesthetics: testing preliminary environmental strategies of the building (orientation, services, etc.) and evaluating their success or otherwise from the perspective of the different users of the building.

   o Cleaning and maintenance: considering how easy the building is to clean and maintain.
5. Energy use analysis, benchmarking and triangulation (see Figure 5.1) of the findings from desktop research, physical monitoring and occupant questionnaires.

Figure 5.1 is an example of a methodological diagram for the post-occupancy analysis of a university library. The elements of study and analysis can be broken down into three primary subjects: energy, environment and occupants, whose interaction is critical to success.

The study elements of in-use BPE are implemented through the following methods of investigation:

- *On-site preparation:* the energy (sub-metering), water and environmental monitoring equipment should be installed during or post-retrofit, but before initial occupation. It is advisable to keep a close eye on the legibility, usability and completeness of monitored data as it is being transferred throughout the project and especially at the outset.
- *Periodic site visits (quarterly, or as needed):* site visits, such as walkthrough surveys and inspections, include observation of user experience, technical reviews of building services equipment, controls and BMS performance. Photographic or video documentation is highly recommended. Walkthrough surveys provide the opportunity for ongoing review and assessment of steps taken post-completion (e.g. fabric analysis, systems installation and commissioning, and control interface review). Furthermore, site visits allow for:
   o questionnaire distribution/collection (e.g. BUS) – timing should be considered; it is necessary to allow the occupants to settle into their new environment and gain experience of it before they can develop opinions

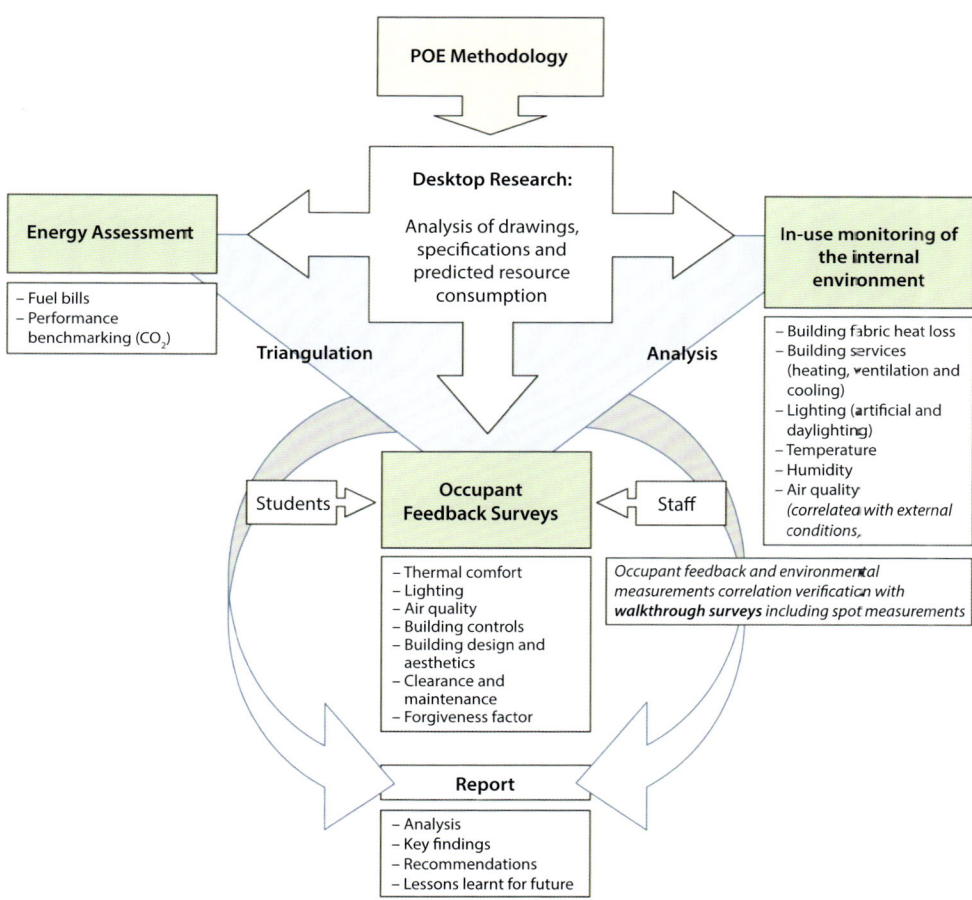

POE Methodology

Desktop Research:

Analysis of drawings, specifications and predicted resource consumption

Energy Assessment

– Fuel bills
– Performance benchmarking (CO_2)

In-use monitoring of the internal environment

– Building fabric heat loss
– Building services (heating, ventilation and cooling)
– Lighting (artificial and daylighting)
– Temperature
– Humidity
– Air quality
  (*correlated with external conditions,*

Triangulation

Analysis

Students

Occupant Feedback Surveys

Staff

– Thermal comfort
– Lighting
– Air quality
– Building controls
– Building design and aesthetics
– Clearance and maintenance
– Forgiveness factor

Occupant feedback and environmental measurements correlation verification with **walkthrough surveys** *including spot measurements*

Report

– Analysis
– Key findings
– Recommendations
– Lessons learnt for future

5.1 Example of a methodological diagram for the BPE study of a university library

How do you find the temperature at this time?		How do you find the lighting level in your workspace at the moment?	
Much too warm		Very bright	
Too warm		Bright	
Comfortably warm		Slightly bright	
Comfortably neither warm nor cool		Neither bright nor dim	
Comfortably cool		Slightly dim	
Too cool		Dim	
Much too cool		Very dim	

5.2 Excerpts from a user satisfaction questionnaire survey

o   semi-structured interviews with occupants and facilities managers

o   spot measurements of environmental variables (e.g. temperature, humidity, etc.)

o   periodic infrared thermography

o   basic checks to ensure that meters are operating and that staff responsible for energy monitoring understand the systems and software (additionally, to check that usage is generally as expected)

o   structured reviews with occupants and management soliciting suggestions for improvement with follow-through and continual verification of changes and resultant reporting of impacts.

Importantly, periodic evaluation allows for investigation of issues, anomalies and/or problems to be caught and resolved quickly. This is knowledge gained not only to solve problems for informed decision-making and localised improvement, but also for future professional development and sector-wide building improvement.

• *Ongoing data collection, analysis, feedback and dissemination (at least one year in duration):* this stage includes periodic (defined based on the duration of the study) analysis of energy, water and environmental variables being monitored along with occupant satisfaction survey data (see Figure 5.2). The process includes:

o   collection of fuel and utility bills to complement metered and sub-metered data

o   assessment of energy use – CIBSE *TM22: Energy Assessment Tool* – a systematic framework for undertaking an energy survey and reporting and benchmarking the results

o   analysis of energy demand profiles – findings from completed profile forms and/or walkthrough discussions with key stakeholders can be an essential method for revealing the 'why' behind the consumption patterns, informing opportunities for fine-tuning performance

o   triangulation of energy use, environmental data and occupant satisfaction/behaviour patterns

o   analysis referring back to post-completion findings indicating relevant issues, such as areas of heat loss and user controllability

o   annually, energy use (and resultant CO_2 emissions) results can be assessed in a meaningful way and benchmarked against predicted performance, comparable buildings and best-practice benchmarks (using CIBSE *TM46: Energy Benchmarking*).

Findings from the in-use BPE are compared with the pre-retrofit baseline, which may be the design targets set at the pre-retrofit stage or the actual pre-retrofit performance of the building, to ascertain the scope and level of energy, CO_2 and cost reductions achieved through retrofitting. In this way, a BPE study can also act as a tool for monitoring and verification of the post-retrofit performance of non-domestic buildings.

CONCLUSION

This essay has laid out a robust and pragmatic process map of BPE for achieving and evaluating low energy and low carbon retrofitting of non-domestic buildings in the UK and beyond. Practically, BPE studies depend on participants' willingness; first, to have their building (including design) examined (pre- and post-retrofit) and, second, to have their own occupancy of the building examined in detail. Scrutiny of design, construction and performance at this level of detail has often been difficult to implement for fear of what might be exposed. Soft Landings, for example, has effectively created a framework for highlighting the attractiveness and benefits of a BPE process that is in place to assist rather than simply expose. This exposure (or rather passing on of lessons learned without blame or negativity), however, is desperately needed for ongoing improvement in the building industry, design profession and facilities management role. After all, it is only a matter of time before BPE becomes mandatory for all new builds and retrofits, which will be required to be low energy and low carbon.

ENDNOTES

1 UKGBC (UK Green Building Council) (2013). *Retrofit*. Available online at: www.ukgbc.org/content/retrofit (accessed 27 November 2013).

2 C. H. Pout and F. MacKenzie, *Reducing Carbon Emissions from Commercial and Public Sector Buildings in UK*, Watford, Building Research Establishment (2005).

3 W. F. E. Preiser and J. Vischer, 'The Evolution of Building Performance Evaluation: An Introduction', in W. Preiser and J. Vischer (eds), *Assessing Building Performance*, Oxford, Elsevier (2005), pp. 3–13.

4 W. F. E. Preiser, H. Z. Rabinowitz and E. T. White, *Post-Occupancy Evaluation* New York, Van Nostrand Reinhold (1988).

5 W. F. E. Preiser and U. Schramm, 'Building Performance Evaluation', in D. Watson, M. J. Crosbie and J. H. Callendar (eds), *Time-Saver Standards: Architectural Design Data*, New York, McGraw-Hill (1997).

6 Preiser and Vischer (2005).

7 A. Kato, P. Le Roux and K. Tsunekawa, 'Building Performance Evaluation in Japan', in W. Preiser and J. Vischer (eds), *Assessing Building Performance*, Oxford, Elsevier (2005), pp. 149–159.

8 S. Mallory-Hill, T. van der Voordt and A. van Dortmont, 'Evaluation of Innovative Workplace Design in the Netherlands', in W. Preiser and J. Vischer (eds), *Assessing Building Performance*, Oxford, Elsevier (2005), pp.149–159.

9 S. Ornstein, C. de Andrade and B. Leite, 'Assessing Brazilian Workplace Performance', in W. Preiser and J. Vischer (eds) *Assessing Building Performance*, Oxford, Elsevier (2005), pp.149–159.

10 J. Carthey, 'Post Occupancy Evaluation: Development of a standardized Methodology for Australian Health Projects', *International Journal of Construction Management*, 2006, pp. 57–74.

11 J. Goins, *Case Study of Kresge Foundation Office Complex*, Center for the Built Environment, University of California, Berkeley (2011). Available online at: http://escholarship.org/uc/item/30h937bh (accessed 27 November 2013).

12 R. Cohen, M. Standeven, B. Bordass and A. Leaman, 'Assessing Building Performance in Use 1: The Probe Process', *Building Research and Information*, 2001, Vol. 29, Issue 2, pp. 85–102.

13 Usable Buildings Trust, *Initial Thoughts on a Programme of Post-occupancy Evaluation*, Usable Buildings Trust's submission to the Department for Business and Regulatory Reform (BERR) (2008).

14 Sea trials involve the final phase of construction where a sea vessel's performance is tested in terms of speed, manoeuvrability, endurance, etc. The sea trial is designed to observe changes at the cutting edge of technological development and to feed forward integration of emergent concepts and technologies. G. England, V. Clark and J. L. Jones, *Naval Transformation Roadmap: Power and Access from the Sea*, Darby PA, Diane Publishing (n.d.).

15 Usable Buildings Trust (UBT) and Building Services Research and Information Association (BSRIA), *The Soft Landings Framework: For Better Briefing, Design, Handover and Building Performance In-use*, BSRIA BG 4/2009 (2009).

16 BREEAM, *BREEAM In-Use* (2009). Available online at: www.breeam.org/page.jsp?id=373 (accessed 27 November 2013).

17 R. Gupta, M. Gregg and R. Cherian, 'Tackling the Performance Gap Between Design Intent and Actual Outcomes of New Low/Zero-Carbon Housing, forthcoming in *ECEEE 2013 Summer Study on Energy Efficiency*, 3–8 June 2013, Presqu'île de Giens, Toulon/Hyères, France.

18 Technology Strategy Board (TSB), *Energy-efficient Whitehall: Competition for Development Contracts* (2010). Available online at: www.innovateuk.org/_assets/pdf/competition-documents/briefs/sbri_energyefficientwhitehall.pdf (accessed 27 November 2013).

19 Technology Strategy Board (TSB) (2012). *Invest in Innovative Refurbishment: Competition for SBRI Contracts* (2012). Available online at: www.innovateuk.org/_assets/0511/sbri_comp_innovative_refurbishment_webfinal.pdf (accessed 27 November 2013).

20 W. F. E. Preiser, 'The Evolution of Post-occupancy Evaluation: Toward Building Performance and Universal Design Evaluation', in *Learning From Our Buildings: A State-of-the-Practice Summary of Post-occupancy Evaluation*, Federal Facilities Council Technical Report No. 145, Washington DC, National Academy Press (2002), pp. 9–22.

21 Usable Buildings Trust (UBT) and Building Services Research and Information Association (BSRIA) (2009).

22 B. Bordass, *Performance Evaluation of Non-domestic Buildings*, presented at the Technology Strategy Board Building Performance Evaluation Call Launch, London, 19 May 2010.

23 C. Zimring, M. Rashid and K. Kampschroer, *Facility Performance Evaluation* (2010). Available online at: www.wbdg.org/resources/fpe.php# (accessed 27 November 2013).

24 Cohen *et al.* (2001).

25 Institute for Sustainability (IfS), *04 Technical appendices* (2011). Available online at: http://bob.instituteforsustainability.org.uk/knowledgebank/public/bpereport/guide-4/Pages/4-1-.aspx (accessed 27 November 2013).

26 Gupta *et al.* (2013).

27 B. Bordass, A. Leaman and R. Bunn, *Controls for End Users: A Guide for Good Design and Implementation*, Reading, Building Controls Industry Association (2007).

28 Gupta *et al.* (2013).

29 Technology Strategy Board (TSB), *Building Performance Evaluation, Non-Domestic Buildings: Technical Guidance* (2011).

ADDITIONAL SOURCES

B. Bordass and A. Leaman, 'Making Feedback and Post-Occupancy Evaluation Routine 1: A Portfolio of Feedback Techniques', *Building Research and Information*, Vol. 33, Issue 4, 2005, pp. 347–352.

B. Bordass and A. Leaman, 'A New Professionalism: Remedy or Fantasy?', *Building Research and Information*, Vol. 41, Issue 1, 2013, pp. 1–7.

B. Bordass, R. Cohen and J. Field, *Energy Performance of Non-Domestic Buildings: Closing the Credibility Gap* (2004). Available online at: www.usablebuildings. co.uk/Pages/Unprotected/ EnPerfNDBuildings.pdf

B. Bordass, R. Cohen, M. Standeven and A. Leaman, 'Assessing Building Performance In Use 3: Energy Performance of the Probe Buildings', *Building Research and Information*, Vol. 29, Issue 2, 2001, pp. 114–128.

CarbonBuzz, *The Performance Gap* (n.d.). Available online at: www.carbonbuzz.org (accessed 27 November 2013).

Carbon Trust, *Closing the Gap: Lessons Learned on Realising the Potential of Low-carbon Building Design*, London, Carbon Trust (2011).

Department of Energy and Climate Change (DECC), *How the Green Deal Will Reflect the In-situ Performance of Energy Efficiency Measures* (2012). Available online at: www.gov.uk/government/uploads/ system/uploads/attachment_data/ file/48407/5505-how-the-green-deal-will-reflect-the-insitu-perfor.pdf (accessed 27 November 2013).

Good Homes Alliance (GHA), *GHA Monitoring Programme 2009–11: Technical Report. Results from Phase 1: Post-construction Testing of a Sample of Highly Sustainable New Homes* (2011). Available online at: www.goodhomes.org.uk/downloads/ pages/GHA%20Monitoring%20 Report%20-%20APPROVED.pdf (accessed 27 November 2013).

R. Gupta and S. Chandiwala, 'Understanding Occupants: Feedback Techniques for Large-scale Low-carbon Domestic Refurbishments', *Building Research and Information*, Vol. 38, Issue 5, 2010, pp. 530–548.

A. Leaman, F. Stevenson and B. Bordass, 'Building Evaluation: Practice and Principles', *Building Research and Information*, Vol. 38, Issue 5, 2010, pp. 564–577.

chapter 6

MARK SIDDALL

NON-DOMESTIC RETROFIT
PROJECTS IN GERMANY AND THE USA

INTRODUCTION

Low energy retrofit is clearly a subject of global relevance. International exchange of knowledge and information is, however, in its very early stages. The German Passivhaus Standard has proved particularly transferable. While in-use data are hard to obtain, there appear to be some high achieving non-domestic retrofit projects in Germany using the Passivhaus Standard. In the USA, the retrofit drive originated in the commercial sector, driven by the impact of energy use on rental yields and perceptions of asset value, leading to a 'quick win' approach to retrofit. While the knowledge may not be wholly applicable to the UK, there are valuable lessons to be learnt from each of the two countries' experience to date.

FACING PAGE Mildmay Community Centre, London. See Case Study 8

71

The availability of robust data relating to the replacement and refurbishment rate of non-domestic buildings is currently limited and the quality is generally poor. Within England and Wales, data from 1998 established that the refurbishment rate of non-domestic buildings, across a range of sectors, lay between 2% and 8%, with the average being just under 3%.[1] Roughly one-third of these British non-domestic buildings stem from before the nineteenth century and a further third from between 1900 and 1960. The replacement rate of non-domestic buildings in the UK is not yet well-established, though it may be higher than that of domestic buildings (which is less than 0.7% per year). In 2004, Hartless noted that anecdotal information suggests that just 1–2% of Europe's total building stock (domestic and non-domestic) is replaced each year.[2] Even if the UK non-domestic rate is two or three times higher than the domestic rate, it is generally thought that over 70% of the buildings that are expected to be standing in 2050 have already been built.

The European Union's Energy Efficiency Directive (EED)[3] requires that each year from 1 January 2014, central governments renovate 3% of the total floor area that they own and occupy. The requirements of the Directive have the potential to promote the greening of public procurement and to prime the construction industry for a step change in energy efficiency of existing buildings and the development of relevant skills and knowledge. The database for Central Government Property and Land including Welsh Ministers Estate[4] offers a snapshot of the public estate as of 1 September 2011:[5]

- the floor area of government buildings is approximately 16.5 million m^2, not including space occupied by local authorities, schools or the NHS
- the estate extends to 13,911 different properties
- approximately one-fifth of the estate lies within London (17.4%)
- 40% of the estate comprises office space; the estate also includes 40 laboratories and 18 museums
- there are 552 vacant properties.

In order to comply with the EED, a 33-year programme will be required, necessitating the refurbishment of 492,350 m^2 of the total estate per year. At say, between £500 and £1,000 per m^2, the annual expenditure would be some £250m to £500m per annum; and that is for just the public estate leaving aside over 3m privately owned non-domestic buildings. So how can we ensure that this vast amount of public money is wisely spent on the retrofitting process? Three types of retrofit project are generally recognised, as detailed below.

1. Operations and maintenance improvements through commissioning and better use of the existing building. Research undertaken in the USA, involving hundreds of existing non-domestic buildings, has determined that the recommissioning of building services alone delivers a median energy saving of 16%.[6]
2. Standard retrofits, which may be described as low-risk efficiency upgrades that are undertaken using incremental capital expenditure in order to provide cost-effective energy savings. Such well-intentioned projects often fail to capitalise on their true potential.
3. Deep retrofits, which have the potential to benefit from the integration of all design measures and engineering systems in order to generate maximum energy savings. Although there is no clear line to define precisely when a retrofit is deep enough, the Passivhaus and EnerPHit Standards, which are discussed in more detail below, offer benchmarks for best practice energy efficiency and consequently demands integrated design methods.

There is a growing number of international new-build projects that have achieved a reduction in heating demand of up to 80% and a reduction in cooling demand of 50% or more as compared to the prevailing norm; there are also examples of similarly impressive reductions in electrical demand (lighting and plug loads). Can similar energy efficiencies be achieved with non-domestic retrofit and, if so, how?

The need to retrofit buildings is a global challenge. The art and science of retrofitting buildings to achieve the high levels of energy efficiency required to mitigate climate change is still in its infancy and there is, as yet, little international exchange of knowledge. The very different climatic conditions around the world make the experience of some countries more relevant to the UK than others. This chapter focuses on two sets of retrofitted buildings: one set in Germany, built using the Passivhaus/EnerPHit Standard, and the other in the USA, where the very high energy demand of commercial offices is triggering a significant retrofit programme.

INTEGRATED DESIGN

Financially, the cumulative cost of individual, incrementally applied measures can be reduced by implementing numerous measures as part of a single overall deep retrofit project. A further benefit of such a strategy is that the size of mechanical plant, and the infrastructure that it requires, can be optimised. On the grounds of energy and maintenance cost savings alone, the initial capital cost of a deep retrofit can be difficult to justify. Combining a deep retrofit with other refurbishment activities can improve the viability of a deep retrofit project; in such cases, building components that have reached the end of their usable life, such as windows, are replaced with new units optimised to suit the building. Experience of integrated whole systems design in the USA typically results in designs that aim to achieve energy savings of 45% or more while in some European countries energy savings of 80% have been targeted.

Therefore integrated design and deep retrofit are fundamentally suited to two scenarios. In the first, a building owner has the opportunity to undertake retrofit measures that can be combined with other refurbishment operations. In the second, a building owner has to achieve ambitious energy-saving goals within a limited period of time as a consequence of external conditions or factors, such as legislation.

STANDARDS

UK standards for retrofit are still in a state of development. For example, CIBSE Guide F has recently introduced a section on retrofit. The somewhat out-of-date ECON 19 for offices suggests that for standard, mechanically ventilated buildings the energy demand is roughly 300–330 kWh/m^2/yr, for good practice is 173–186 kWh/m^2/yr and that good practice, naturally ventilated buildings are achieving about 127–145 kWh/m^2/yr.

Whatever the picture with regard to UK standards, looking at good practice more widely is a powerful guide to design of retrofit projects.

GERMANY

Germany has run a number of programmes focusing on new build performance targets. Programmes such as these can be used as a means of developing an appreciation of the huge potential for exemplary retrofit. Two programmes in particular stand out:

1. *Solarbau*: the Solarbau programme[7] targeted a primary energy demand of 100 kWh/m^2/yr (the current conversion factor for primary to delivered electricity is approximately 0.4; for directly burned fuels it is 1). In total, 23 projects were undertaken. A number of office buildings used the Passivhaus Standard as a means of focusing the design.
2. *Research for Energy Optimised Construction (EnOB)*: this programme has examined advanced construction targeting <60 kWh/m^2/yr of end-use energy, representing a 50% reduction compared to 2007 building standards (as with Solarbau this equated to <100 kWh/m^2/yr primary energy). Detailed analysis of 11 new build projects was undertaken by Kalz et al.[8] The measured energy consumption ranged between 18 kWh/m^2/yr and 65 kWh/m^2/yr; some 4 to 7 times less than the average new office or institutional building. The buildings all exploit thermally-activated building systems (TABS), heat sources and heat sinks. In two cases, office buildings recover exhaust heat from the

server room and use it to supply heat via TABS. More recently, EnOB has also engaged with retrofits, the goal being to improve upon the 2007 EnEv Standard (the German energy code) by some 30%. A total of 17 case studies have been undertaken or are currently in progress; at this time only limited data are available for many of the schemes.

In addition to advanced standards being investigated through Government-funded programmes, there is also the adoption of the voluntary quality assurance standards, such as Passivhaus and EnerPHit. Many regard the Passivhaus Standard as the world's leading quality assurance standard for energy-efficient construction. Passivhaus design may be characterised as low energy design that focuses on delivering excellent standards of comfort during both the summer and the winter; these comfort standards will be explored in more detail below. It is this focus on addressing thermal comfort that sets Passivhaus apart from a straightforward energy efficiency standard. The way in which the standard is defined makes it particularly transferable to international applications.

The standard was first established in 1996 following detailed research between 1989 and 1991 by Dr Wolfgang Feist and Professor Bo Adamson. It was originally for domestic application; however, it has proven to be appropriate for a growing range of non-domestic buildings. The word 'haus' actually means something more akin to 'building' (e.g. Kunsthaus means art gallery). There are over 50,000 Passivhaus buildings, a number of which are non-domestic and include offices, schools, laboratories, swimming pools, fire stations, hotels, court houses, art galleries and, more recently, hospitals. There is also a growing number of non-domestic retrofits. In recognition of challenges specific to retrofit, the Passivhaus Institute launched the EnerPHit Standard in 2011. The headline design requirements for the Passivhaus and EnerPHit Standards are detailed in Table 6.1. There are also several sub-criteria that inform and support these main headings, given in the Passivhaus Planning Package.

Table 6.1 Passivhaus and EnerPHit Standards

Performance requirement	Passivhaus performance criteria[9]	EnerPHit performance criteria[10]	Rationale
Airtightness	≤0.6 ac/h @ 50 Pa	≤1.0 ac/h @ 50 Pa	Comfort and wellbeing
Surface temperature of windows	≥17 °C on design day	≥17 °C	
Ventilation	~ 30 m³/h per person	~ 30 m³/h per person	
Supply air temperature	≥16 °C	≥16 °C	
Summer overheating	≤25 °C for ≤10% per annum	≤25 °C for ≤10% per annum	
Design temperature	20 °C	20 °C	
Specific heat demand	≤15 kWh/m²/yr ≤10 W/m²/yr	≤25 kWh/m²/yr	Energy
Primary energy	≤120 kWh/m²/yr*	≤120 kWh/m²/yr	

* Depending on building type, the primary energy demand for non-domestic buildings may vary.

Energy definitions in Passivhaus

- Primary energy: a measure of the energy at the point where it was generated.
- Primary energy includes energy lost during the generation and transmission of electricity.
- Delivered energy: a measure of the energy at the point where it was utilised.
- End-use energy: does not include energy lost during the generation and transmission of electricity.

As Table 6.1 makes clear, a number of key criteria focus upon thermal comfort: air temperature, radiant temperature, draught risk, etc. In the Passivhaus philosophy, energy demand is tackled by addressing wellbeing and comfort criteria via passive and low energy design. The secret of success in this approach lies in the integration of needs with wellbeing, comfort and low energy objectives.

In recognition of the constraints imposed by existing buildings, the specific heat demand target of the EnerPHit Standard may be satisfied if Passivhaus design measures are utilised (i.e. maximum U-values of 0.15 W/m²K for opaque elements and 0.8 W/m²K for windows, doors, etc.). This and further requirements are set out in the document *Certification Criteria for Energy Retrofits with Passive House Components*.[11]

In order to deliver deep retrofits, the Passivhaus and EnerPHit Standards require an integrated design approach.

CASE STUDIES
OFFICE BUILDING, 41–43 WERNER-VON-SIEMENS STRASSE, ERLANGEN[12]

Constructed in 1972, this six-storey building, with an additional two-storey basement, was typical of its time and included 50 mm of internal insulation and a heater/air-conditioning unit that was integrated within the external wall below aluminium ribbon windows. Occupants experienced functional problems with the controls for the heating/cooling units; for instance, during the summer, if the room air-conditioning mode cooled the room too much, the heating could then switch on. Furthermore, the building services had reached the end of their functional life and needed to be replaced.

As computer usage rose during the 1990s, energy use also increased; resulting in an additional transformer having to be installed to cover peak loads and a requirement for 30 additional cooling units.

By 2006, the building's owner/occupier Daeschler decided that a holistic retrofit was required in order to future-proof the building and the design team

6.1 Care and attention is given to not only the external envelope, but also the energy demand within the building

was appointed. The building refurbishment programme retrofitted Passivhaus components in order to minimise energy demand.

Measures

The aluminium windows were replaced by windows with 2+1 glazing that contains integral blinds/reflectors (2+1 windows utilise double glazing with an additional external pane of glazing – the cavity between the double and single glazing can be used to facilitate solar shading). External thermal insulation was used to improve fabric U-values by a factor of between five and ten: 0.17–0.19 W/m^{2}K for walls, 0.17–0.21 W/m^{2}K for roofs, 0.24 W/m^{2}K for the basement and 1.0–1.3 W/m^{2}K for the windows. The air-conditioning system was replaced by one with highly efficient heat recovery and radiant ceiling panels that provide both heating and cooling. While creating an environment that is practically draught-free, these changes allowed the supply air volume to be reduced from 250,000 m^3/h to just 20,000 m^3/h with subsequent savings in fan power, pipe and duct diameter and ventilation heat losses.

The majority of the power used by the building is provided by the co-generation unit. The heat that is generated is used to supply space heating during the winter and, via an adsorption chiller, cooling in the summer. This system is supported by a number of buffer tanks consisting of an ice tank, cool tanks and a warm tank. Weather compensation is integrated into the controller so that surges in demand can be minimised. Cooling loads have been further reduced via the night cooling of the storage tanks; thus saving electricity while also increasing system efficiency. The cooling demand, which would formerly use 100 kW of peak-rate electricity during the daytime, is therefore reduced to a mere 10 kW. Pipe sizing has been optimised for minimal power use by circulation pumps.

Energy performance

The above measures were calculated to reduce the primary energy demand of the building from 600 kWh/m^2/yr to just 100 kWh/m^2/yr with co-generation providing 65 kWh/m^2/yr to achieve this figure.

The work was completed for €830 per m^2, covering 11,475 m^2 of gross floor area and 8,150 m^2 of net floor area. The building services were subject to a monitoring and commissioning programme. Consultation with the building users was undertaken to help optimise performance.

CLIENT: Daeschler
ARCHITECT DESIGN: Werner Haase Architects
ARCHITECT IMPLEMENTATION: Scheerzer Architects
SERVICES ENGINEER: Beck and Trommen
ENERGY CONSULTANT: Werner Haase Architects/Beck and Trommen
CONTRACTOR: Bauherr

ELEMENTARY SCHOOL, BAIERSDORF [13–15]

Undertaken in three phases between 2005 and 2006, this project included partial demolition and partial replacement/extension. Built in 1959, the school had undergone numerous changes and adaptations. The old boilers provided about 530 kW of heat and resulted in energy bills in the region of €29,500 per year.

The basic concept for the retrofit design was that the heat from the 350 students, 21 teachers, lighting and equipment would cover 90% of the school's space heating requirements; only at external temperatures below -4 °C would space heating be required. In principle, the heating is switched off once the school opens as it is then heated by the activities within the building. The schoolchildren have been estimated to raise the indoor temperature by 2 °C.

Measures

The school was retrofitted using Passivhaus components. Fabric U-values were substantially improved with 200 mm of external thermal insulation applied to the walls to give U-values of 0.15–0.19 W/m^{2}K; 200 mm cellulose insulation with a further 60 mm wood fibre insulation was applied to the main roof, and 160 mm of mineral wool insulation was installed on the roof of the halls giving U-values of 0.17–0.21 W/m^{2}K. There was limited scope for improving the U-value of the floors. Existing timber windows were replaced with 2+1 windows with a U-value 0.95 W/m²K; containing integral blinds. Overheating risks within the building are minimised through the use of night purge ventilation and solar shading.

In order to minimise ventilation heat losses, the airtightness of the building was improved and decentralised heat recovery units were installed, with each unit serving two classrooms. Measurements taken in the existing building revealed indoor CO_2 levels of 3,000 ppm. Supply air, at a flow rate of 15–20 m³/person, was used to limit indoor CO_2 levels to no more than 1,200–1,500 ppm. Measurement of indoor CO_2 levels after refurbishment found that they did not exceed 900 ppm. Typically, the units achieve two air changes per hour, though they are run at 450 m³/h for half an hour prior to school commencing and then switched off out of hours. Intake and exhaust ductwork passes through the external walls. This ventilation strategy avoided the need for a large central unit, fire dampers and sound proofing. All units are equally sized and of the same type to keep maintenance as simple as possible.

In order to reduce the quantity of primary energy that is used to satisfy the space heating demand, a ground source heat pump with 12, 9 m deep, 500 mm diameter boreholes was specified.

Radiant heating panels are integrated within the walls and thermostats are provided to enable localised control. The supply temperature of the panels ranges from 22 °C to 29 °C.

In addition to the heat pumps, there is a gas-powered combined heat and power (CHP) system, with an output of 5.5 kW electricity and 14.5 kW heat, and a gas-fired condensing boiler has been provided to offer top-up heat as required. Five 1,000 litre buffer vessels have been provided to balance heat supply diurnally from the CHP.

6.2 (top) The use of external insulation served to reduce the risk of thermal bridging. (bottom) The intake and exhaust vents for each classroom can be seen on the facade. Passivhaus components have been used to minimise heat loss. The intake and exhaust ductwork are visible

Energy performance

The design achieves a space heating demand of 50 kWh/m²/yr – a 72% reduction in demand compared to the 185 kWh/m²/yr of the school prior to retrofit. Primary energy of 50 kWh/m²/yr equates to about 19 kWh/m²/yr delivered energy.

The construction cost of the retrofit was €980/m², covering 3,032 m² of treated floor area.

CLIENT: Baiersdorf
ARCHITECT: Werner Haase Architects
ENERGY CONSULTANT: Werner Haase Architects
CONTRACTOR: Bauherr

OFFICE, MANUFACTURING AND WAREHOUSE[16, 17]

Built in 1959, this former Telefunken building, used for manufacturing televisions, served as accommodation for a printing company between 1980 and 1998. Purchased in 2008 by Hanover-based AS Solar, it had stood empty for ten years. The existing structure (partially of three and partially of two storeys) is rectangular: 94 m x 53.5 m, with a height of 12.1 m. There is also a 6.2 m deep basement. The retrofit project entailed stripping the building back to the existing reinforced concrete structure.

The refurbished building provides accommodation for the 250 staff of AS Solar with 7,000 m² of office space in addition to the manufacturing and storage facilities. The energy efficiency of the building was set zone by zone, based on the operational temperature within each zone. The operational temperature within office areas, and similar spaces, was set at 20 °C. This proportion of the building was refurbished to satisfy the Passivhaus Standard; the 1,000 m² of manufacturing space was refurbished to German Energy Savings Regulations 2009 and the remainder, the storage space, was left unheated. The retrofit measures were informed by the use of the Passivhaus Planning Package (PHPP) and dynamic simulation, taking into account internal gains, was used to assess overheating risks and plan how the building would be used. Building performance is being monitored by the University of Braunschweig.

Measures

Walls were fitted with prefabricated panels containing 240 mm of cellulose insulation with a further 60 mm of wood-fibre insulation that then received an external render finish. The target U-value was 0.12 W/m²K.

Optimising the glazing area for daylight and solar gains strongly influenced the design. Curtain walling with a U-value of 0.8 W/m²K was used to provide openable windows and natural daylight. Four rooflights were introduced to improve daylighting within the office areas. Diffusers were introduced to assist with the distribution of daylight and reduce the risk of glare.

The existing flat roof received 450 mm of cellulose insulation between the timber framing and was then overlaid with trapezoidal sheeting. Photovoltaic panels were affixed to pitch roof elements formed by the frame. The target U-value for the roof was 0.11 W/m²K.

The basement was upgraded to achieve a U-value of 0.21 W/m²K.

The building was sealed to reduce air leakage to 0.6 per hour at 50 Pa.

The heat recovery ventilation system, with a rate that ranges between 6,600 m³/h and 15,300 m³/h, serves the bistro, foyer, seminar rooms, kitchen and offices. Fresh air is introduced via suspended ceiling radiant panels. Fire dampers were installed where required to achieve fire compartmentation and separation. Openable windows provide cooling during the summer. The manufacturing area is naturally ventilated all year and heated by radiant heating panels. Fresh air is pre-warmed or cooled to 20 °C depending on requirements.

Space heating is provided via a number of biomass boilers and solar thermal panels, which have been selected to showcase products manufactured by

AS Solar. Cooling within the offices is provided by adsorption refrigeration powered by solar thermal; 2 x 19 kW cooling power with an additional 160 kW of compression refrigeration to address peak loads. Chilled water is stored reusing an old 2 m^3 compressed air tank that was already installed within the existing building. Heating and cooling within the office space is supplied via radiant heating panels mounted at ceiling level.

The existing 30 m^3 sprinkler tank is now being used as a buffer tank. The tank operates as a central heat store and is linked to both the biomass boilers and the 150 m^2 solar thermal array. Excess heat is exported to the district heating system; monitoring has determined that the heat generated is sufficient to serve 80 homes.

An array of photovoltaic panels (286 kW peak on the roof and 126 kW peak in the car park) is predicted to generate 250,000 kWh/yr.

Electrical lighting is controlled by passive infrared (PIR) detectors and photocells that adjust the lighting levels relative to the level of daylight. Dimmable floor lamps provide lighting. These are supplemented by task lighting to all desk spaces.

Energy performance

As the building has stood empty for ten years, the energy performance pre-retrofit is neither available nor of significance, with the focus of the project being on absolute rather than relative performance. The target space heating demand of the building is the standard 15 kWh/m^2 of treated floor area. The overall primary energy demand is 71 kWh/m^2 (equivalent to approximately 26 kWh/m^2/yr) with CO_2 emissions of 17 kg/m^2. These figures, if sustained in use, would be impressive for a new building, let alone a retrofit. It has not been possible to access in-use energy figures.

CLIENT: AS Solar
ARCHITECT: John M. Frank
ENERGY CONSULTANT: Prof. Dr Lars Kuhl, energydesign braunschweig GmbH

6.3 External, rather than internal, solar shading is used to minimise overheating

BRÜDERSTRASSE, LEIPZIG[18, 19]

This 2,474 m² concrete-framed building of four storeys, a single-storey annex and basement was constructed in 1984. After 20 years of use, doors, windows, interior walls and finishes had reached the end of their useful life and were in need of replacement; consequently, the building was almost completely gutted. The refurbishment included the creation of a daycare centre and education facilities.

The building's compact form was assessed to be beneficial as it enabled the thickness of the insulation to be minimised; thus also minimising capital cost. The modular and standardised components of the existing building were also deemed to be advantageous, as their uniformity enabled external insulation to be applied to the facades in a straightforward repetitive manner. Given the large number of prefabricated concrete buildings that have been constructed throughout Germany, this project is potentially of wide significance.

The existing structure was more than adequate for accepting the changes in building load that were to arise as a consequence of the refurbishment and retrofit. The renovation, to EnerPHit standards of energy performance, commenced in 2010.

The project brief, developed by the Free State of Saxony, stipulated that the budget for the Passivhaus retrofit must cost no more than 8% extra compared to a conventional retrofit to 2004 German Energy Efficiency Guidelines. In order to remain within budget, careful coordination of the design and construction was required. Scenario analysis was undertaken to determine the optimal economic and environmental impact of various options that satisfied the performance requirements. Surprisingly perhaps, conventional technologies and materials were found to offer the greatest value. Financial payback was calculated to be achievable within about 16 years. The building was constructed for €1,004 per m² with a total construction cost of €2.485 m.

Measures

The U-value of the building envelope of 0.12 W/m²K was achieved using 240 mm of expanded polystyrene or, for areas requiring fire protection, 300 mm of mineral wool. Thermally broken windows, achieving a U-value of 0.8 W/m²K, were selected and an EPDM membrane was specified to seal between the frame and the building fabric. Solar shading was provided to the daycare centre.

Particular design challenges arose as a consequence of the existing external envelope of the building being constructed from a prefabricated concrete system. Initial pressure tests of the ground floor storey revealed that air leakage was 1.4 air changes per hour (ac/h) @ 50 Pa, mainly due to cracks in panels, gaps at panel joints and between panels and the ground floor/ basement structures. In order to achieve the targeted airtightness of 0.6 ac/h @50 Pa, an internal air barrier was specified.

The retrofit for the new uses required new fire protection, safety and accident prevention systems. It proved challenging to find appropriate components and products that would also achieve Passivhaus Certification; for instance, the laminated safety glass degrades the thermal performance of the windows to some extent. In order to address the impact that this would have on the design, compensatory measures were applied to certain other aspects. Examples of such measures include upgrading from argon to krypton fill within the glazing, improving the U-values or the efficiency of the heat recovery unit, or applying a combination of all three measures.

The two separate purposes for the retrofitted building necessitated zoning the heating and hot water systems. Within each zone, different standards of daylighting, shading and ventilation have also informed the design.

Heat, for space, domestic hot water and ventilation, is primarily provided by the existing district heating system with additional energy provided by a solar thermal array upon the roof. Underfloor heating was used on the ground floor with radiators on upper floors. All heating systems were designed to operate at low temperatures.

Cooling is provided by an adsorption chiller, which gets most of its heat from the solar thermal array. Additional heat can also be provided by the district heating system when necessary.

A heat recovery system has been installed with an efficiency of 80%. Openable windows are used to provide ventilation outside the heating season.

Energy performance

The design targets a space heating demand of 17.2 kWh/m²/yr – slightly below the Passivhaus Standard. The total primary energy demand is projected as 57.5 kWh/m²/yr – equivalent to delivered energy of approximately 21 kWh/m²/yr.

USA

The USA covers a wide range of climate zones, some of which are unfamiliar within a European context. Consequently, the direct relevance of these case studies to European practice may be limited. However, developing an appreciation of how practice can be adapted to different or extreme climates can present unexpected gems of knowledge and insight that could prove useful.

Within the USA, commercial buildings are reported to consume 35% of the nation's electricity and 13% of the nation's natural gas.[20] With non-domestic buildings imposing such significant demands upon these resources, low energy retrofit is generating interest, discussion and a growing body of research. While there are no clearly defined retrofit standards for non-domestic buildings in the USA, 'roadmaps' and guidance notes exist, as do benchmarks for completed projects such as those set out in Tables 6.2 and 6.3 (see overleaf).

The figures for buildings in cold climates (the best equivalent to the UK's climate) are indicated in bold. They show that, while a number of US retrofit projects are achieving high percentage reductions in energy relative to pre-retrofit demand, the absolute numbers are much higher than equivalent UK and European best practice.

6.4 (left) The nursery within the Bruderstrasse retrofit. (right) Triple glazed windows reduce the risk of thermal discomfort arising due to downdraft and low radiant temperatures

Table 6.2 Offices[21]

	Site energy use intensity* (EUI) kWh/m²/yr		Site (EUI)* reduction (%)	Financial analysis			
	Baseline	Post-deep retrofit		Total package cost ($)	Total annual savings ($)	Simple payback	Net present value ($)
Hot and humid	278	151	45	697,000	117,000	6.0	227,000
Hot and dry	306	145	52	890,000	161,000	5.5	422,000
Marine	297	139	53	918,000	162,000	5.7	369,000
Cold	**271**	**139**	**48**	**885,000**	**153,000**	**5.8**	**302,000**
Very cold	287	148	49	837,000	137,000	6.1	211,000
Average	287	145	50	845,000	146,000	5.8	306,000

Average area 18,581 m²

Table 6.3 Schools[22]

	Site energy use intensity* (EUI) kWh/m²/yr		Site (EUI)* reduction (%)	Financial analysis			
	Baseline	Post-deep retrofit		Total package cost ($)	Total annual savings ($)	Simple payback	Net present value ($)
Hot and humid	265	205	23	829,394	109,516	7.6	901,098
Hot and dry	271	196	28	1,165,461	118,949	9.8	764,048
Marine	259	177	32	1,022,096	170,000	6.0	1,744,787
Cold	**325**	**221**	**32**	**1,089,271**	**288,944**	**3.8**	**3,577,302**
Very cold	391	265	32	877,669	120,479	7.3	1,126,919
Average	303	211	30	996,778	161,578	7	1,622,831

CASE STUDIES

The conviction that a high environmental rating is a marketable commodity in the property business has led a number of property companies to invest in energy efficiency upgrades of office properties. Glenborough LLC was among the first. The company's corporate retrofit policy contains five categories:

- energy efficiency
- water conservation
- waste management
- tenant education and
- procurement and operational best practice.

'The Aventine' at La Jolla, California is a LEED Platinum Certified retrofit, which also achieves the highest grade in the US Environment Protection Agency's voluntary Energy Star scheme. Before deciding to retrofit The Aventine, originally designed by Michael Graves, Glenborough retrofitted 1525 Wilson Boulevard in Roslyn, Virginia. These two buildings together provide an interesting snapshot of evolving retrofit best practice in the USA.

1525 WILSON BOULEVARD, ROSLYN, VA[23, 24]

Constructed in 1987, this 12-storey 29,110 m^2 building is located on 'The Hill', one of the strongest suburban office markets in the USA. The structure consists of ground floor retail, storage and a three-storey basement car park, with the remainder being offices. Before the retrofit, the building was a mid-market office with 100% tenancy; however, the energy demand represented one of highest in the Glenborough portfolio. All of the building systems are powered by electricity.

The tenancy mix consists of government agencies and their contractors and large institutional firms.

6.5 1525 Wilson Boulevard

One of the aims of the project was to secure the building's future success by raising the office to the top of the market through improving efficiency, comfort and sustainability.

The trigger for undertaking the retrofit stemmed from the need for major upgrades and replacement of the aging heating, ventilation and air-conditioning (HVAC) systems. Economic analysis suggested that the most cost-effective strategy would be to improve the efficiency of equipment and systems that were subject to high loads and high demand. Therefore, air distribution systems and HVAC compressors (responsible for 50–60% of the building's energy use) were prioritised, along with an upgrade of lighting systems and reduction in plug loads.

The greatest challenge of the project was retrofitting the building while preventing interruptions to tenant working conditions and ensuring that the building remained fully occupied throughout the works. This was achieved by undertaking work at night-time and at weekends. In addition, efforts were made to engage with tenants in order to explain the benefits that the retrofit would offer and to establish suitable feedback mechanisms.

The energy efficiency upgrade, which was undertaken within a one-year programme, resulted in an energy demand reduction of 35%; saving $250,000 and reducing CO_2 emissions by 1,250 metric tons.

During the retrofit, a number of existing VAV (variable air volume) units were found to be disconnected, not maintained or, in some cases, undocumented; the result being that whole zones of the building were not adequately serviced. The retrofit package addressed the VAV system and included the retrofit of 90% of the compressors and sensors. An open source control system was utilised to allow future upgrades to be made and next-generation technology to be integrated.

Envelope

A review of the theoretical performance of the fabric was undertaken and it was determined that no further action need be taken to improve the thermal performance of the envelope. Neither a thermographic survey nor an air leakage test were used to assess the building envelope, though gaskets to doors and windows were inspected.

Had the air leakage been determined, then equipment could have been sized more appropriately, and both comfort and cost efficiencies improved; particularly if the envelope had undergone remediation prior to the retrofit of the HVAC system. However, upgrading the fabric would have caused disruption and loss of rental income. This illustrates perfectly a key problem in commercial retrofit – passive measures, which offer the best long-term solution to energy efficiency, are the most problematic in the short term.

Lighting

High wattage lamps were replaced by lower wattage lamps and fixtures. High pressure sodium lamps within garage areas, which run 24 hours a day, were replaced by LED lights. Within office areas, T8 fixtures were replaced by T5 ballasts and lamps.

Management

Glenborough negotiated a day-time cleaning programme so that the building did not require interior illumination at night.

Controls

The control of compressors was converted from pneumatic controls to Tridium direct digital controls (DDC) – an open-source framework that permits next-generation technology to be integrated with greater ease. Lighting control was provided by the inclusion of occupancy sensors and time controls.

Commissioning

All commissioning was undertaken by Glenborough's engineers.

Monitoring

An Energy Information System is being trialled in order to examine a range of loads that had not been monitored previously. In the future, Glenborough intends to implement metering in all tenant spaces, as well as a programme that will assist tenants in reducing plug loads.

Execution plan

Glenborough has found that easy access to historical energy data is vital. This data enables the company to understand the energy demand within the building and to plan and prioritise its energy management strategy by targeting opportunities for energy efficiency and energy saving. In addition to the energy management strategy, a clear business case for the energy efficiency and energy saving activities must be made. The business case should include a comprehensive review of revenue and expenses, incorporating:

- market conditions, asset ownership period and ownership entity
- tenant lease expiration and projected voids in tenancy (occupancy levels)
- tenant leases with recoverable operational or capital expenditures.

Tenant engagement

Other buildings in the company's portfolio, such as The Aventine, show that plug loads represent 7.5–15% of the energy demand. Glenborough considers that it is vital to educate and engage with its tenants so that these loads may be reduced. Such reduction also helps to reduce the cooling loads that are imposed on the HVAC system.

Energy performance

In 2008, the energy demand of 1525 Wilson Boulevard, excluding all energy associated with the plug loads, was 309 kWh/m²/yr. The Commercial Building Energy Consumption Survey (CBECS[25]), based on 4,500 properties, has found that average energy demand of US Office Buildings is 203 kWh/m²/yr. The database from which Table 6.2 for offices is produced states a benchmark for 'Very cold' climates as 287 kWh/m²/yr. Relative to either benchmark, the Wilson Boulevard pre-retrofit figure was high. By 2010, the building was demanding 202 kWh/m²/yr of delivered energy, excluding all energy associated with plug loads. This is substantially better than average, although it falls well short of the average deep retrofit benchmark shown in the table (145 kWh/m²/yr).

Cost and funding

Excluding labour costs for Glenborough maintenance staff, the project cost was $1,100,000. As a consequence of the economic downturn, financing was provided from Glenborough's own internal funds.

The simple payback period for the measures has been estimated to be two years. This is an example of a modest retrofit; in this case, only the HVAC plant yielded a considerable saving in energy and cost.

Estimated energy savings are 2,300,000 kWh of electricity with a value of $283,000 at a low rate of 7.5 cents per kWh. There are also maintenance savings in the order of $75,000.

CLIENT: Glenborough LLC
MECHANICAL ENGINEER: Alpha Mechanical
GENERAL CONTRACTOR: Alpha Mechanical
PHOTOGRAPHY: Alan Schindler

THE AVENTINE, LA JOLLA, CA[26]

Designed by Michael Graves and completed in 1990, The Aventine is a class 'A' office building. The building comprises an 11-storey structure with an adjacent six-storey wing and three levels of basement parking. While The Aventine was 100% occupied at the time of the retrofit, the market within which the building competed was subject to early terminations and low occupancy levels. A key goal of the retrofit was to improve competitiveness, and maintain 100% tenancy, by reducing operational costs. For this ambition to be achieved, a strong business case for undertaking a cost-effective retrofit was critical; particularly as the building at that time was less than 16 years old.

Critical aspects of Glenborough's retrofit strategy were to generate interest in the retrofit by communicating its advantages to tenants and to avoid disrupting them while the work was being carried out. Consultation events were used to disseminate information to tenants. Retrofit activities were undertaken at night and during weekends. High-energy loads were reduced by replacing compressors and installing chillers, lighting and controls.

HVAC

As mentioned earlier in connection with 1525 Wilson Boulevard, Glenborough had established that around half the energy use in its office buildings arose from the HVAC systems. Despite the fact that the chiller plant had recently been upgraded, the design team determined that optimisation of the chiller plant would still be worthwhile.

A feasibility study showed that the greatest energy savings for the least financial outlay could be achieved by upgrading to variable speed chiller plant with automated controls, while retaining two existing centrifugal chillers. The simple payback period was calculated to be less than three years. In order to minimise the use of fan power and maximise indoor air quality, CO_2 sensors were employed and linked to the control system so that fan speeds and air supply would be modulated to suit demand.

Lighting

The wattage of the exterior lighting was reduced by 75% and internal lighting was improved by switching to compact fluorescent lamps and sensor controls.

Envelope

An air leakage test is reported to have been undertaken, although the specific results are not available. A design review of the 16-year-old envelope determined that no significant upgrades were necessary. A 'cool roof' with high reflectance (albedo), as advocated by the Environment Protection Agency, was installed in order to help reduce cooling loads and improve energy efficiency.

Finances and incentives

The project cost was $801,540 before rebates. Additional internal costs for labour are not available. The utility company provided $175,000 in incentives for upgrading HVAC and lighting.

In 2004, Glenborough targeted a payback of two to three years; however, by 2008, providing that the long-term asset value was increased, retrofit strategies became more aggressive and a payback period of three to four years was targeted. This has more recently been extended to four to six years on the basis that the measures deliver increased long-term asset value and lead to favourable improvements, attracting more desirable tenants.

User satisfaction

It is reported that both user satisfaction and productivity have increased.

Tenant education

Regular outreach events are held in order to engage with tenants and provide them with fun, tangible examples that can help them to reduce energy and resource demand.

Competitive positioning within the market

Reasons given for the building's competitive advantage include:

1. lower operating costs for tenants
2. operations and maintenance programmes which result in fewer equipment failures and lower operating and replacement costs
3. better indoor air quality from systems and controls using state-of-the-art and/or next-generation technology.

Energy performance

Prior to the retrofit, The Aventine demanded 196 kWh/m²/yr of energy. This is 7 kWh/m²/yr below the Commercial Buildings Energy Consumption Survey (CBECS) average of 203 kWh/m²/yr. However, it is some 57 kWh/m²/yr above the 139 kWh/m²/yr good practice threshold for a marine environment.[27]

Following the retrofit, Glenborough reported that the energy demand for The Aventine had been reduced to 73 kWh/m²/yr (excluding all energy associated with plug loads). Compared to the pre-retrofit measurements, there has been a 63% reduction in energy demand, taking it to some 66 kWh/m²/yr below the US good practice threshold for such environments.

During the first year of post-retrofit use, the project is said to have saved over 2 million kWh of electricity, reduced energy and operating expenses by more than $116,000 and reduced carbon emissions by roughly 272 tons.

While the project appears to offer a picture-postcard image of retrofits within the USA, it is worth recognising that the definition of a marine environment spans from the relatively mild Californian climate (the location of The Aventine) up to Seattle. Given the range of latitudes that this category covers, some questions may be raised about the size of the climate zones that have been used to identify good practice within the USA.

Given the starting point, the achievements of this retrofit are certainly worthy of recognition and serve to demonstrate how a commercial landlord may continue to develop and refine their approach towards the retrofit of new buildings within their portfolio. That such a significant improvement in relative performance can be achieved on a building that was just 16 years old says as much about the quality of the existing building as it does about the absolute quality of the retrofit measures that were undertaken. When asset values, improvements to longevity of tenancy and the reduction in voids between tenancies are considered, then extending the duration of the payback period is a pragmatic development of a business plan. What this begins to indicate is that the more evolved the business model, the deeper the potential for energy-efficient retrofit.

OWNER: Glenborough, LLC
ORIGINAL ARCHITECT: Michael Graves Architects
MECHANICAL ENGINEER: Glenborough LLC

6.5 The Aventine

CONCLUSIONS

Assessing retrofit performance is highly dependent on how boundary conditions are set on different projects. A target saving of 100 kWh/m²/yr in one project may not be comparable to 100 kWh/m²/yr in another project; as is particularly the case when comparing projects in Europe and the USA. There are many ways of assessing energy, such as considering specific energy, delivered energy and primary energy; and there are different methodologies for measuring floor area. Furthermore, percentage improvements are relative and can be misleading; absolute numbers are therefore preferable. However, this information is not consistently accessible. For these reasons, most numerical performance data require interpretation.

In Germany, as in the UK, the energy-efficient retrofit started in the domestic sector, driven by regulations, and is now well established. Non-domestic retrofits are increasingly following the Passivhaus methodology, which originally matured in the domestic sector. As the case studies show, energy use in these deep retrofit projects is being reduced to a level close to that of comparable new buildings. However, in-use data are still difficult to source and it is not clear whether the performance gap, so often noted in the UK, is as prevalent in Germany.

In the USA, the energy efficiency drive originated in commercial offices and has been fuelled by cost savings. Because of the far higher levels of energy use, in relative terms, especially in buildings with obsolete plant, large energy savings have been achieved for relatively little capital outlay. In absolute terms, however, the energy or emission level reductions that most of these retrofit projects achieve are far short of what climate science suggests is required. Once the 'low-hanging fruit' of simple plant upgrade has been picked, the gathering momentum of the energy efficiency drive makes it likely that deep retrofit projects will also be designed and implemented. The voluntary benchmarking provided by the Environmental Protection Agency's 'Energy Star' scheme has already played a striking role in this otherwise private-sector-driven transformation and will continue to do so.

A report by the USA-based Rocky Mountain Institute states that there is 'a growing body of statistical evidence [which] suggests that green office buildings can command rent premiums of 3–6 percent and sales price premiums of 10 percent or more'.[28] The business model that has been developed by Glenborough LLC is, in comparison to many others, fairly sophisticated and considers not only saved energy and operating costs, and capital cost avoidance over the life of improvements, but also takes into account rental rates, improved asset values and improved tenancy profiles. As shown in the earlier case studies, their business model has been extended to consider payback periods of between four years (25% return on investment, ROI) and six years (16% ROI); meanwhile, 200 Market Associates in Portland, Oregon will consider payback periods of up to seven years (14% ROI).[29] Such business models are just a beginning. It is conceivable that this kind of model could also be developed to consider reduced financial and regulatory risk and, through working with tenants and occupiers, it could also include, to some degree, improved employee health and wellbeing, reduced absenteeism and improved productivity. Understanding retrofits within this broader context brings a fresh appreciation of the value that can be derived from such activities and, consequently, enables a reappraisal of the risks that such projects may entail. If commercial landlords and owner-occupiers continue to develop new ways of thinking about their buildings, then even more refined business models may evolve. New business models of this kind have the potential to open doors to even more radical reductions in energy demand.

ENDNOTES

1   C. H. Pout, S. A. Moss and P. J. Davidson, *Non-Domestic Building Energy Fact File*, BRE Report 339 (1998).

2   R. Hartless, *ENPER-TEBUC Project Final Report of Task B4 Energy Performance of Buildings: Application of Energy Performance Regulations to Existing Buildings* (2004). Available online at: www.seattle.gov/environment/documents/enper_b4.pdf (accessed 29 November 2013).

3   *Directive 2012/27/EU of The European Parliament and of the Council, Official Journal of the European Union.* Available online at: http://eur-lex.europa.eu/LexUriServ/LexUriServ.do?uri=OJ:L:2012:315:0001:0056:EN:PDF (accessed 29 November 2013).

4   *Central Government Property and Land Including Welsh Ministers Estate.* Available online at: http://data.gov.uk/dataset/epims (accessed 29 November 2013)

5   Press release, *Government saves more than £100 million on property this financial year*, Cabinet Office, 12 January 2012. Available online at: www.gov.uk/government/news/government-saves-more-than-100-million-on-property-this-financial-year (accessed 29 November 2013).

6   E. Mills, *Building Commissioning: A Golden Opportunity for Reducing Energy Costs and Greenhouse Gas Emissions*, Lawrence Berkeley National Laboratory (2009).

7   Available online at: www.solarbau.de/english_version/index.htm (accessed 29 November 2013).

8   D. E. Kalz, S. Herkel and A. Wagner, 'The Impact of Auxiliary Energy on the Efficiency of the Heating and Cooling System: Monitoring of Low-energy Buildings', *Energy and Buildings*, Vol. 41, 2009, pp 1019–1030.

9   W. Feist, *Criteria for Non-residential Passive House Buildings*, 25 April 2012. Available online at: http://passiv.de/downloads/03_certfication_criteria_nonresidential_en.pdf (accessed 29 November 2013).

10 W. Feist, *Certification Criteria for Energy Retrofits with Passive House Components*, 25 April 2012. Available online at: http://passiv.de/downloads/03_enerphit_criteria_en.pdf (accessed 29 November 2013).

11 *Ibid.*

12 W. Haase *Complete Renovation of the Office Building at Werner-von-Siemens-Strasse 41–43 in Erlangen*, 12th International Conference Proceedings on Passive Houses, 2008, Nuremberg.

13 G. Lommer, *Festschrift, Einweihung der Grundschule Baiersdorf* (2007).

14 Personal notes by author from Passivhaus Tour 2008

15 *Energieeffizientes Bauen in Bayern.* Available online at: www.regierung.mittelfranken.bayern.de/ (accessed 29 November 2013).

16 Pro Clima, *Passivhaus Diversity*, Excursions Brochure, 16th International Passivhaus Conference, Hanover, 2012.

17 L. Kuhl, S. Ludwig, P. Eickmeyer, M. Schlosser and T. Wilken, AS Solar, Hanover – existing plus-energy industrial building, Proceedings of the 16th International Passivhaus Conference, 2012.

18 C. Kodderitzch and A. Matuschak, *Specific Requirements and Problems with the Refurbishment of a Non-Residential Pre-Fabricated Building to Passive House Standard*, Proceedings of the 14th International Passivhaus Conference, 2010.

19 Universität Leipzig, *Vom Plattenbau zum Passivhaus – Energieeffizienzmaßnahme Brüderstraße 16 in Leipzig*. Available online at: www.sib.sachsen.de/uploads/media/79_10_Universitaet_Leipzig_Abschluss_Energieeffizienzmassnahme.doc (accessed 29 November 2013).

20 D. Moser, G. Lui, W. Wang and J. Zhang, 'Achieving Deep Energy Savings in Existing Buildings Through Integrated Design', *ASHRAE Transactions*, Vol. 118, 2012, pp. 3–10.

21 G. Liu *et al.*, *Advanced Energy Retrofit Guide: Practical Ways to Improve Energy Performance: Office Buildings*, Pacific Northwest National Laboratory, September 2011. Available online at: www.pnnl.gov/main/publications/external/technical_reports/pnnl-20761.pdf (accessed 29 November 2013).

22 G. Liu *et al.*, *Advanced Energy Retrofit Guide Practical Ways to Improve Energy Performance: School Buildings*, Pacific Northwest National Laboratory, February 2013.

23 New Buildings Institute, *Deep Energy Savings in Existing Buildings Case Study: 1525 Wilson Boulevard* (2011). Available online at: www.newbuildings.org/sites/default/files/Case_Study_1525-Wilson-Blvd.pdf (accessed 29 November 2013).

24 New Buildings Institute, *A Search for Deep Energy Savings in Existing Buildings* (2012). Available online at: www.josre.org/wp-content/uploads/2012/10/11Deep-SavingsEBCaseStudiesNBI.pdf (accessed 29 November 2013).

25 US Energy Information Administration, Commercial Buildings Energy Consumption Survey (CBECS).

26 New Buildings Institute, *Deep Energy Savings in Existing Buildings Case Study: The Aventine* (2011). Available online at: www.newbuildings.org/sites/default/files/Case_Study_Aventine.pdf and www.josre.org/wp-content/uploads/2012/10/11DeepSavingsEBCas-eStudiesNBI.pdf (accessed 29 November 2013).

27 P. Torcellini, S. Pless, M. Deru, B. Griffith, N. Long and R. Judkoff, *Lessons Learned from Case Studies of Six High-Performance Buildings*, National Renewable Energy Laboratory, Technical Report NREL/TP-550-37542 (2006).

28 S. Muldavin, 'Beyond the Tip of the
    Energy Iceberg: Why Retrofits Create
    More Value Than You Think', *Solutions
    Journal*, Summer 2013, Vol. 6,
    No. 1. Available online at: www.rmi.org/
    summer_2013_esj_beyond_the_tip_of_
    the_energy_iceberg_main (accessed
    29 November 2013).

29 Better Bricks, NEEA, *Existing Building
    Renewal Case Study: 200 Market
    Building*. Available online at: www.
    betterbricks.com/sites/default/files/
    Design%20%26%20Construction/bb_
    casestudy_200market_final.pdf (accessed
    29 November 2013).

CASE STUDIES

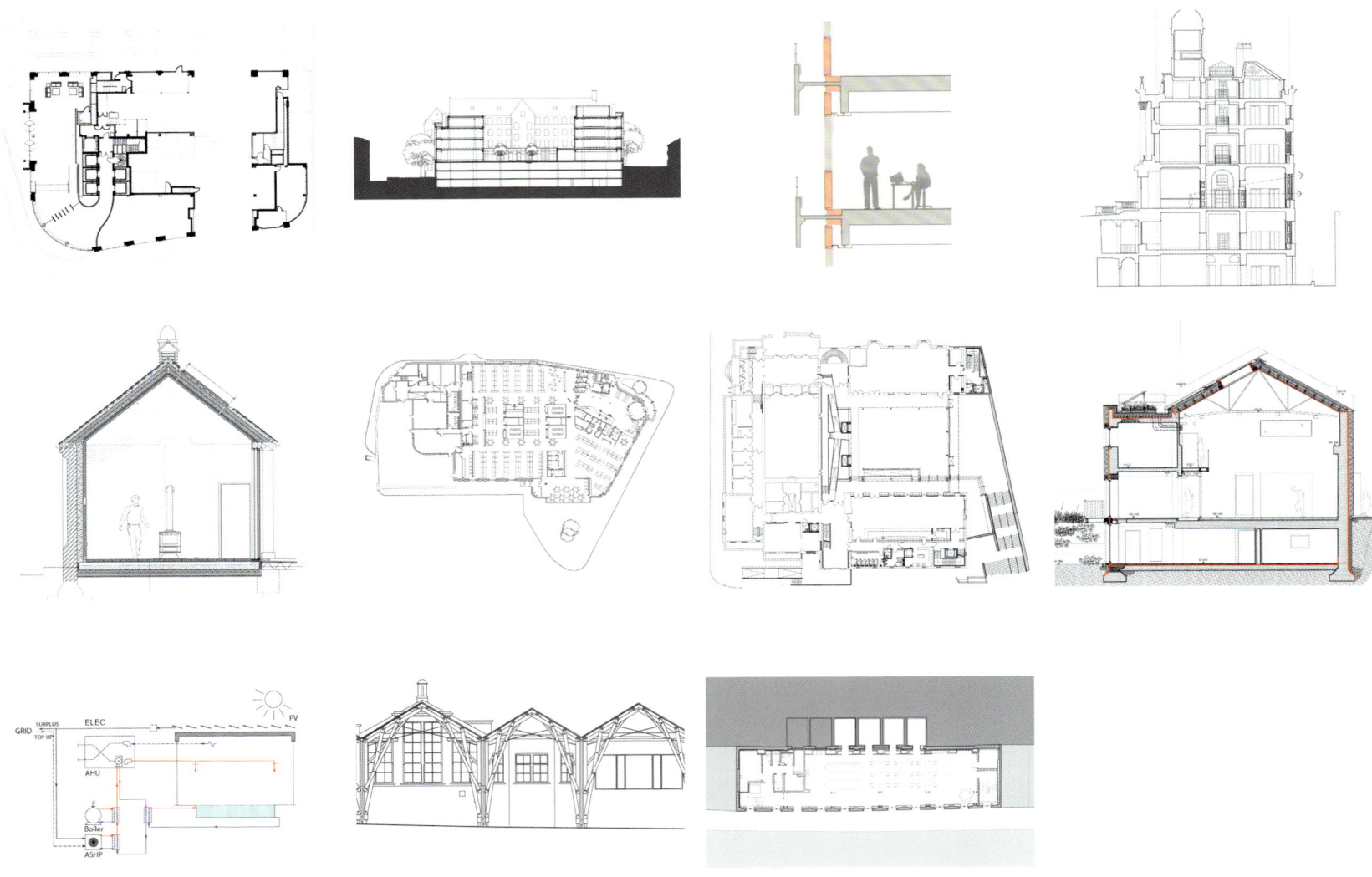

The 11 case studies in the following pages are a representative rather than a comprehensive selection of the best completed retrofit projects in the UK of recent years. They were selected by looking at refurbishment awards entries and invitations via the *Architects' Journal*, and by contacting individuals and groups such as the UK Green Buildings Council and the Edge, the multidisciplinary built-environment professional think tank. The wish to include case studies of many building types, both in terms of use and period, itself acted as a filter, limiting the numbers of any one type. 'Retrofit' covers a wide spectrum – all the way from the minimal, such as overhauling the mechanical plant to the maximal, where the building is stripped right back to the frame and the form is totally reconfigured, so that it becomes almost like building anew. A recent example of the latter is the excellent, Stirling Prize shortlisted Angel Building, designed by Allford Hall Monaghan Morris for Derwent Properties, which is not included here. We have stayed in the middle of the spectrum in selecting the case studies; the one exception being Elizabeth II Court, designed by Bennetts Associates for Hampshire County Council. The building housed HCC offices before refurbishment and so the project illustrates a wide-ranging asset consolidation strategy, returning large reductions in energy use.

The most challenging element in compiling the case studies has been the assembly of energy and carbon data in a consistent format, so that performance can be compared both 'before' and 'after' (where post-occupancy data are available) and also across various buildings in terms of actual energy use. Even in the best of projects there are discontinuities in design and construction teams, which mean that there is no one person in custody of all the data. At the start of the project, pre-retrofit data may not have been collected in sufficient detail to enable a full analysis of post-retrofit performance – a lack of data about plug loads being the most common gap. During design the parameters may have changed, with earlier ambitions for large energy savings being diluted when cost implications fully kicked in. Once the project is complete there may be insufficient time (or will) to doggedly collect and analyse performance data, especially since they tend to be disappointing! Nevertheless, the clients and designers of the buildings illustrated here have been willing to share the performance figures – and that is one of the most important things for the art of retrofit today.

Both energy consumption and carbon emissions are included in the energy tables presented in the case studies. Many different conversion factors are in use today to derive CO_2 emissions from energy, and they tend to change from year to year as the carbon intensity of various fuel mixes changes. The factors are also commonly expressed to three, four or even five significant figures, which is of little use in retrofit, where the actual performance typically varies by a factor of 2 plus. The Carbon Trust's current (2013) numbers are 0.18404 $kgCO_2e$/kWh for gas and 0.44548 $kgCO_2e$/kWh for grid electricity. For Building Regulations calculations, the numbers in common use are 0.198 and 0.517 $kgCO_2$/kWh respectively. The Passivhaus method, now called Passive House in the UK, has its own more stringent 0.68 $kgCO_2$/kWh for electricity. We have decided to simplify the whole thing, and escape yearly adjustment cycles by applying an easy to remember 0.2 $kgCO_2$/kWh for gas and 0.5 $kgCO_2$/kWh for electricity. The energy figures are always present so that readers can apply their own factors if they wish to in order to make comparisons with other projects.

In writing these succinct case studies we have tried to remain true to the overriding philosophy that energy efficient design and architectural design need to be seen as an integrated whole; that creating uplifting settings for life and saving energy should be of mutual value.

NOTE
CO_2e stands for 'equivalent carbon dioxide', which takes into account the global warming potential of all main greenhouse gases. The case studies adopt the Building Regulations approach of referring only to $kgCO_2$

Level 11
Level 10
Level 09
Level 08

Level 07
Level 06
Level 05
Level 04

Level 03
Level 02
Level 01

01

199 Bishopsgate, London

BRITISH LAND PLC

This project offers many lessons and sets a benchmark for the energy-efficient retrofit of air-conditioned office buildings constructed in the boom of the late 1980s, which are now coming up for refurbishment at the end of their first 25-year lifecycle. While delivering commercial value without compromise – the ultimate test being the attractiveness to prestigious tenants – the retrofit has better integrated the building with its urban context, approximately halved predicted energy use and reduced predicted emissions by 40%, while also achieving other sustainability aims. Notably, these measures are complemented by a programme of engagement with the occupiers, to help them save operational energy.

01 New entrance and lobby
02 Corner entrance and lobby pre-retrofit

CONTEXT AND OBJECTIVES

Completed in 1988, the original 11-storey building was designed by Skidmore, Owings & Merrill as Phase 14 of the Broadgate development. Built over the railway tracks serving Liverpool Street Station, it occupies a key corner site at the junction of Bishopsgate and Primrose Street, is overlooked by a conservation area and lies within a protected view of St Paul's Cathedral.

While the building was structurally sound, changes in technology, communications and working practices since its completion meant that it no longer met the demands of current and future occupiers. Rather than demolish and rebuild, Bluebutton Properties UK Ltd, a joint venture between British Land and Blackstone, took the decision to modernise and refurbish the building. There were three reasons for this decision:

- to improve the quality of the space – the opportunity to add floor area was limited as the existing building already filled the available footprint and was the maximum height allowed by a strategic view corridor. Instead, it was decided to improve the quality of the space by reducing the net internal area, to be achieved through enlarging the reception and removing a mezzanine at the top of the building
- to reduce carbon emissions, both operational and embodied
- to position the building as 'new' in the letting market, showcasing the client's commitment to sustainable development – the refurbishment was to dramatically improve energy performance, office environment, facilities and materials.

The brief was that the building should bring 199 Bishopsgate back into the Broadgate 'family' by means of delivering office space finished to Category

A standard and benchmarked against the British Council for Offices' best. The building had to be capable of being let as a whole or on a floor-by-floor basis. The key objectives of the retrofit were to:

- upgrade the quality, flexibility and layout of the office space
- improve the layout of the building entrance and enhance the arrival experience to the building
- create 'active' frontages on Bishopsgate and Primrose Street
- improve the energy performance of the building, targeting a BREEAM 'Excellent' rating
- implement a new building services strategy to meet modern and future standards
- bring the building up to, in the words of British Land, 'world-class office standards, optimising the attractiveness for occupiers in the global professional services, finance and insurance, and technology, media and telecommunications sectors'
- to extend the life of the building fabric and services by 15 years before the next major maintenance.

Side by side with the commitment to sustainability, the client required a rigorous approach to the cost-effectiveness of proposed solutions.

DESIGN: GENERAL

Externally, the most visible aspect of the redesign is a new base to the building – this orients it four-square to Bishopsgate, the reason for the original diagonal orientation having been superseded by the expansion of developments northwards. Behind the bigger scale of glazing that forms this two-storey base is a far grander lobby, of a restrained classical corporate design, with the now more generous lift bank reached by a route that uses the previous chamfered facet to sweep you round the corner. It is difficult to tell that this is not a brand new building. Looking up the building, from the second floor on there is no visible change until the three top floors, which have been totally remodelled to accommodate a new central plant and breakout space on the 11th floor with fine views over the city fringe.

The lettable floors have been made much more flexible in terms of subdivision by the rearrangement of the cores. They have the usual Category A fit-out, leaving the tenants to complete the interior to their own requirements. The floorplates also allow improved views across Shoreditch and towards the Olympic Park.

The finished quality is noticeably good and with a high level of specification, aided by the extensive use of mock-ups and benchmark areas. Crucially, these mock-ups were constructed early within the programme, to allow any changes in design and construction method to be incorporated prior to the manufacture of the individual building elements.

03 From Bishopsgate post-retrofit
04 Route through to security and lifts
05 General view with original corner entrance pre-retrofit
06 General view post-retrofit

05

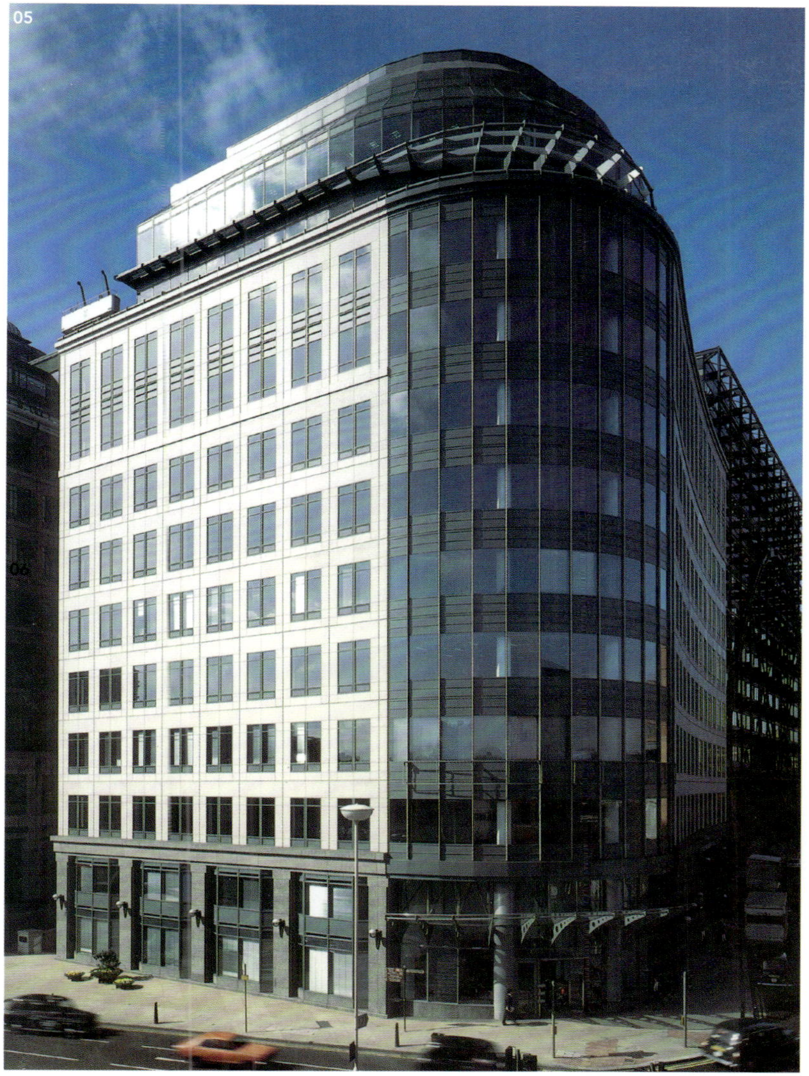

06

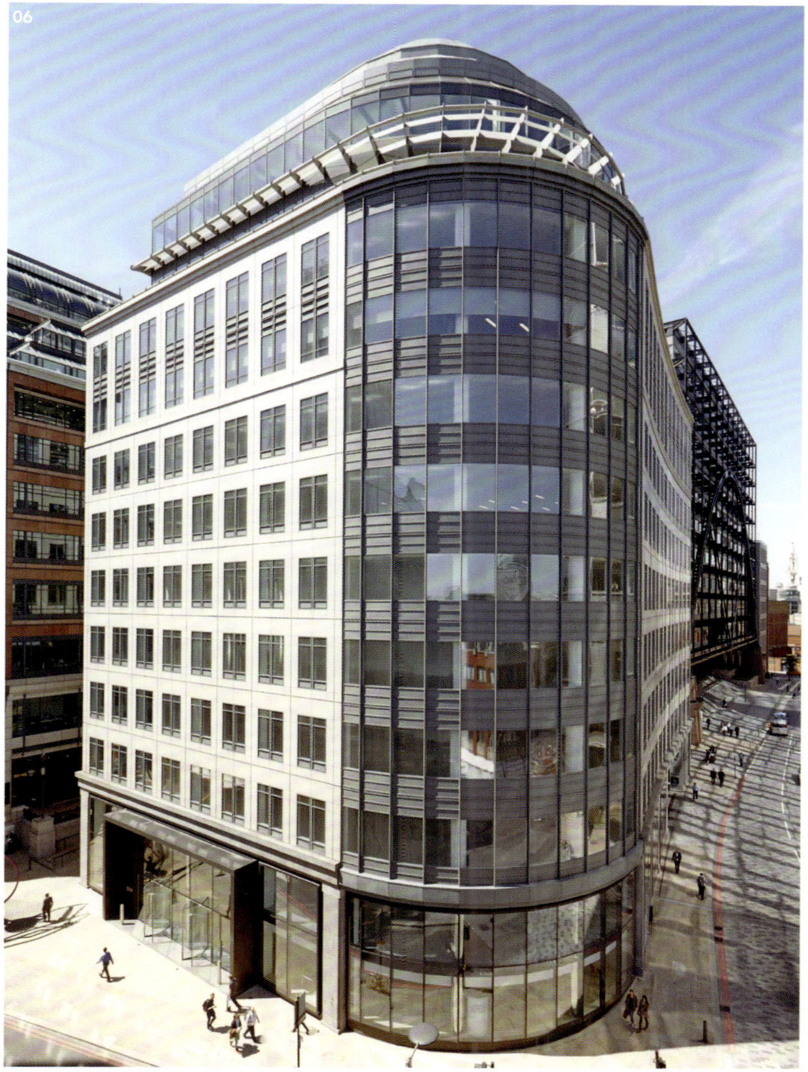

While the building may not wear its environmental credentials on its sleeve, the marketing material for 199 Bishopsgate and British Land's corporate policy make clear that sustainability is a differentiator for the company at two levels – it sets British Land apart from the competition, and it attracts a different kind of occupier, one that is more likely to be there for the longer term.

DESIGN: ENERGY CONSERVATION

The energy strategy of this retrofit can be understood by looking at how the Energy Performance Certificate rating was changed from 'E', with a score of 93, to the target rating 'B', with a score of 47. The cooling load is about 20% higher than the heating load, but because of the different carbon intensities of gas (used for heating) and electricity (used for cooling), the carbon impact of cooling is almost four times bigger than for heating. Thermal modelling showed that redesigning the entire HVAC system and centralising the plant reduced the score from 93 to 60. The use of low-energy lighting and introduction of lighting controls took this down to 50. Other measures included changes to glass specification to reduce solar gain, and some small areas of additional insulation. In contrast to most of the other case studies in this book, very little work was done to the envelope. The only change was that the spandrels below the window cills, where the previous ventilation units were located, received new insulation. The thermal modelling showed that generally reducing the U-value of the envelope would have marginal, if any, benefit,

and the work required would not be cost-effective. An explanation for this apparently anomalous result is that in a building containing equipment producing high internal heat gains, a less-insulated skin allows the heat to escape, with the reduction in cooling load offsetting the increase in the heating load in winter (see also the Guy's Tower case study).

In addition to these retrofit measures, the distribution pipework has been designed to accept future connection to the district heating and cooling system that is already in place in the City of London and expected to be extended. As well as street-level terminals, future connections have been incorporated in the service spine, floor by floor.

09

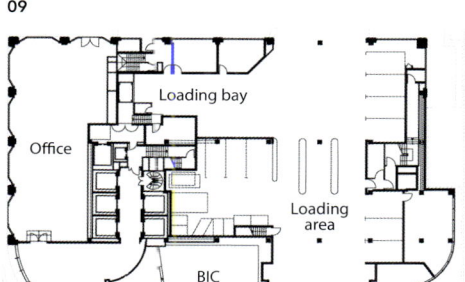

Loading bay

Office

Loading area

BIC

10

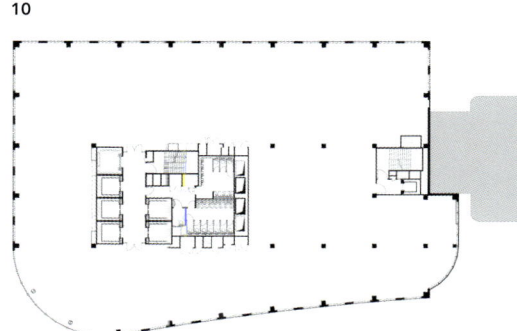

11

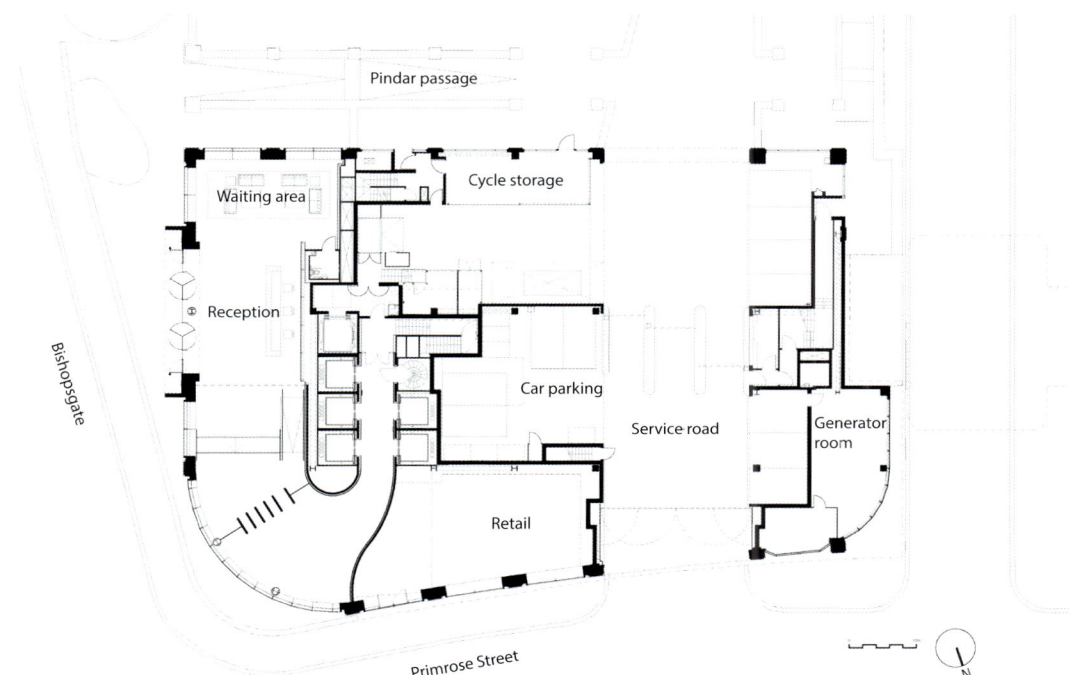

Pindar passage

Cycle storage

Waiting area

Reception

Car parking

Service road

Generator room

Retail

Bishopsgate

Primrose Street

N

07 Bishopsgate frontage pre-retrofit
08 Bishopsgate frontage post-retrofit, showing new entrance
09 Ground floor plan pre-retrofit
10 Typical floor plan
11 Ground floor plan post-retrofit
12 Pre-retrofit plant arrangement
13 Post-retrofit plant arrangement

12

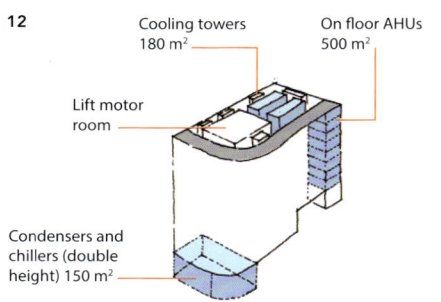

Cooling towers 180 m²

On floor AHUs 500 m²

Lift motor room

Condensers and chillers (double height) 150 m²

13

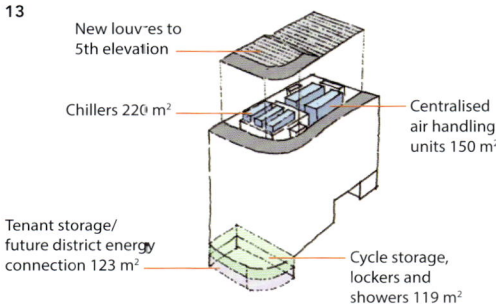

New louvres to 5th elevation

Chillers 220 m²

Centralised air handling units 150 m²

Tenant storage/ future district energy connection 123 m²

Cycle storage, lockers and showers 119 m²

USER BEHAVIOUR AND BUILDING MANAGEMENT

In a project like this, the building's occupants are not known in advance and therefore cannot be recruited in advance to play a part in conserving energy. Instead, British Land operates a proactive method to influence tenant behaviour. The starting point is the Occupant Energy Estimator, which, through sub-metering, enables occupants to gain a greater understanding of the impact of their use of the building and how that correlates with energy and carbon usage. Occupants are then more likely to adjust their behaviour accordingly. The landlord also sets up an Environmental Working Group in each building, providing a forum for beneficial exchange of knowledge between owner and occupier.

Broadgate Estates Ltd (BEL) manages the building on behalf of the landlord, provides specialist maintenance across Broadgate and ensures that the building as a whole will continue to perform to the standards that have been established. BEL audited the RIBA Work Stage E design to ensure the project was 'designed for management', and it participated in the quality inspections at handover. The process of commissioning and environmental testing applied at completion, to ensure the mechanical plant is performing at its optimum efficiency, is to be repeated seasonally.

DATA

Energy and carbon

199 BISHOPSGATE
Commercial tenanted office building

		Energy kWh/m²/yr	Carbon kg CO_2/ m²/yr	Area (m²)	Occupancy (m²/person)	Hours of operation
PRE-RETROFIT	Gas	17.79	3.56			
Estimated	Electricity regulated	121.33	60.67			
2006 carbon figures	Total regulated	139.12	64.22			
	Electricity unregulated	186.72	93.36			
	Total (est.)	325.84	157.58			
POST-RETROFIT DESIGN DATA				18,529 GIA		Office hrs
	Part L model SBEM		13.70	13,600 Net	n/a	
	Predicted energy model (not Part L model)					
	Gas	27.86	5.57			
	Electricity lighting	37.17	18.59			
	Electricity VAC	28.41	14.21			
	Renewables	0.00	0.00			
	Total regulated	93.44	38.36			
	Lifts	6.46	3.23			
	Kitchens/catering	pending occupier				
	Electricity comms room	pending occupier				
	Electricity equipment	63.54	31.77			
	Total	163.44	73.36			
ACTUAL DATA	Post-occupancy data	n/a				
	Gas					
	Electricity lighting					
	Electricity VAC					
	Renewables:					
	Total regulated					
	Electricity					
	Electricity unregulated					
	Total					

BENCHMARKS				Env. rating	EPC rating
				BREEAM 'Excellent'	B

CONVERSION FACTORS	Gas	0.200
kgCO_2/kWh	Electricity	0.500

AIR PERMEABILITY	9.84 m³/hr/m² @ 50pa

Note
1. Conversion factors originally were 0.194 for gas and 0.422 for electricity

The building achieved an Energy Performance Certificate rating of 'B', the pre-refurbishment rating having been 'E'.

The building achieved a BREEAM 'Excellent' rating post-refurbishment.

Resource use and waste

British Land's sustainability brief required that a percentage of materials by value must be recycled materials. A forecast using the WRAP Net Waste Tool (NWT) was undertaken by the team. The NWT, geared towards new-build projects, was used primarily to track performance on recycled content rather than waste, as British Land's key performance indicators for waste already set a precedent. The exercise identified 'quick wins', i.e. the most significant opportunities to improve the recycled content (RC) of the project by using products with good practice levels of RC. As an example, plasterboard selected for walls had an RC of 78.5%, better than the 'standard' RC reported by NWT of 36% and approaching 'good' at 84%. The overall performance for the project was 16% RC, equivalent to a NWT 'good' performance (16%).

The main contractor established a waste recycling centre on site. The resulting substantial diversion from landfill bettered the BREEAM targets and the client's corporate requirements: 97% against 95% saved from landfill for construction/fit-out and 99.97% against 97% saved from landfill for demolition/strip-out.

Energy – achieving 44.49 $kgCO_2$ per £100,000 spend
– exceeds target of minimum 75–100 $kgCO_2$ per £100,000 spend.

Water – achieving 6.59 m^3 per £100,000 spend.

Timber – 100% sustainable timber.

Cost and time

Construction cost:

Total construction cost:	£22.6m = £1,220 per m^2
Demolition and external works:	£1m = £55 per m^2
Shell and core:	£16.9m = £915 per m^2
Category A fit-out:	£4.7m = £255 per m^2

Time from project inception to completion:

Contract duration:	18 months

Procurement

The project was procured under a design and build contract (JCT Major Project Construction Contract 2009). John Robertson Architects had the opportunity to monitor contract compliance and, more importantly, design quality during procurement and construction.

The project was procured through a two-stage design and build tender, with quality assured by developing the employer's requirements to RIBA Work Stage E, thus defining the level of finish and allowing the contractor to include for the expected standard when pricing the job.

14

14 Waiting area facing south

CREDITS

DEVELOPER
British Land and Blackstone

MAIN CONTRACTOR
Como

PROJECT MANAGER
M3 Consulting

COST CONSULTANT
Sense

PLANNING CONSULTANT
DP9

SUSTAINABILITY CONSULTANT
Environmental Perspectives

ARCHITECT
John Robertson Architects

STRUCTURAL ENGINEER
Meinhardt

BUILDING SERVICES ENGINEER
Chapman Bathurst

CDM CO-ORDINATOR
Capita Symonds

STRIP-OUT CONTRACTOR
H Smith (Engineers) Limited

PHOTOGRAPHY
Richard Leeney

Elizabeth II Court, Winchester

HAMPSHIRE COUNTY COUNCIL

In this project, a huge, energy-hungry, functionally outdated 1960s monolith of a building, lodged uncomfortably in a fine city grain, was transformed into an uplifting and adaptable workplace with excellent environmental performance and improved urban connectivity. Combined with the rethinking of working patterns and consequent increase in space utilisation the retrofit has saved the council both operational and accommodation costs and improved workplace morale.

CONTEXT AND OBJECTIVES

Winchester is one of the UK's best loved historic cities, with a rich urban grain and an abundance of significant buildings and spaces. Diverse architectural styles coexist in a relaxed way through an underlying harmony in scale, articulation and quality of materials and construction. Ashburton Court, as it was then named, was built at the height of municipal modernist optimism. It stood in sharp contrast to its context and suffered many of the failings associated with the period. Built around a huge rectangular court dedicated to car parking on two levels, it had a domineering external presence; internally, its working conditions were unsatisfactory and institutional; its energy use was profligate.

For the council the retrofit project was part of an overall strategy of modernising and rationalising its operations. One element of this was to reduce the number of buildings it occupied and to create a better physical relationship with the city via a welcoming and porous interface. Another was to create flexible workspace and a range of new facilities, such as informal work and social places, in line with emerging office design. The project also presented an opportunity to create a benchmark for addressing a contemporary challenge – the creative reuse of 1960s building stock.

DESIGN: GENERAL

Following a thorough feasibility study to compare the costs of replacement, repair and refurbishment, it was decided to embark on an ambitious project that would combine the transformation of the building fabric with a move to flexible working for the staff. The study found that the council could not afford to demolish and replace the building, but that retaining and refurbishing the envelope represented poor long-term value. The recommendation to strip the building back to the concrete frame and implement with an external reclad

01 New café social/work space
02 Bridges in courtyard pre-retrofit

and interior refit was adopted by the council as the best asset management decision. Combined with the removal of 250 staff car parking spaces at ground level (although a semi-basement car park was retained), the transformation enabled the floor area to be increased and the office areas designed to facilitate an open-plan, flexible working environment. The strategy required a large-scale decanting operation, but a programme of more than 2,000 individual moves over a number of phases enabled the council's operation to be maintained with little disruption.

Stripping the building back to the concrete frame enabled the introduction of greater articulation into the form and skyline, as well as higher quality materials that better registered with the surroundings. The new street elevations of paired bays relate to the character, rhythm and materiality of the historic Winchester townscape. Roof-level devices named 'wind troughs' create an articulated skyline that relates to the roofscapes of the city. The integration of town planning and sustainability objectives in a single solution was developed in dialogue with local residents and amenity groups at the planning stage.

Architecturally, the most striking aspect of the retrofit is the design of the new ground plane, formerly a car park. This plane is almost level with the street at its southern end, where the main entrance is located, and a storey up at the north end, along Tower Place. Before the retrofit, enclosed bridges in a T shape hovered above the cars, linking the offices on floors 1 to 6, planned rigidly around a central corridor. The bridges

have been removed and the ground floor designed around two courtyards, either side of a timber-clad lecture theatre. Visitors enter a simple entrance foyer which is flooded with daylight from the external facade and the courtyard. From here, the lecture theatre and the café – located in the long west wing – can be seen across the courtyard, which itself forms part of the well-used set of social spaces of the building. The spatial structure running through these areas could not be simpler, structured by the orthogonal geometry of the frame. The secret of its success lies in the quality of finishes and detailing, creating a relaxed classical feel better than in many expensive corporate offices.

Increasingly, people are working and meeting in these spaces, as well as at their desks, and the 'café' – which is a shared work and meeting space – tends to be packed during the week. The northern half of the west wing has a number of bookable meeting rooms – a key component of the kit of parts enabling new and flexible ways of working. The retrofitted building accommodates 75% more staff than before (1,100 compared with 625 previously). The council's overall office space utilisation has increased by 30%, leading to a 4,500 m² reduction in the total requirement. Alongside the creation of more flexible spaces, the modern IT infrastructure is expected to create further opportunities for efficiency and intensification of use.

Post-occupancy evaluation using Building User Studies methodology found that occupants scored the refurbished Elizabeth II Court highly on design, occupant needs and image.

DESIGN: ENERGY CONSERVATION

The site is surrounded on three sides by heavily trafficked roads, precluding a reliance on opening windows, and the tight floor-to-floor heights prevented the use of low-energy displacement ventilation. The retrofitted building is predominantly naturally ventilated, using an innovative system developed through sophisticated environmental engineering. Acoustic studies indicated that having openable windows was not feasible on the street-facing elevations due to traffic noise levels, so a system was devised whereby air is drawn from the internal courtyards, across the floorplates and expelled through ducts along the street facades. 'Wind troughs' on top of the ducts use wind energy to create the suction force that drives the system, regardless of wind direction. The new street elevations self-shade the building and break up the massing of the facade in a way that relates to the character and materiality of the historic Winchester townscape.

Notably, this project includes no on-site renewable energy generation, concentrating entirely on energy efficiency. Solar shading, intelligent lighting systems that switch off when not required, exposure of the concrete soffits, with their high thermal mass, and a new energy-efficient building envelope all contribute to achieving very significant energy savings. Waste heat from the cooling plant required to service the

03 View of Elizabeth II Court pre-retrofit
04 View of Elizabeth II Court post-retrofit

03

06

04

council's data centre is recycled to heat areas of the building in winter. Water-saving devices in toilets and washrooms keep consumption within previous levels despite the increase in occupancy.

USER BEHAVIOUR
AND BUILDING MANAGEMENT

The relative simplicity of the heating and cooling strategies, together with a degree of automation in the controls, reduces reliance on the occupants' understanding of the operations of the systems and their active engagement in energy conservation.

The culture change most significant for energy conservation has been the introduction of flexible working, which has resulted in increased space utilisation. Hampshire County Council now has more than 50% of its Winchester headquarters staff working flexibly. Together with the increase of the floor area, this has allowed around 500 more staff to be accommodated in the building, which has enabled the council to reduce its central headquarters estate by about 30%.

The design achieved a BREEAM Offices (2006) rating of 'Excellent', with a score of 72.89%.

05 New café social/work space
06 Work environment pre-retrofit
07 Work environment post-retrofit
08 External facade post-retrofit

05

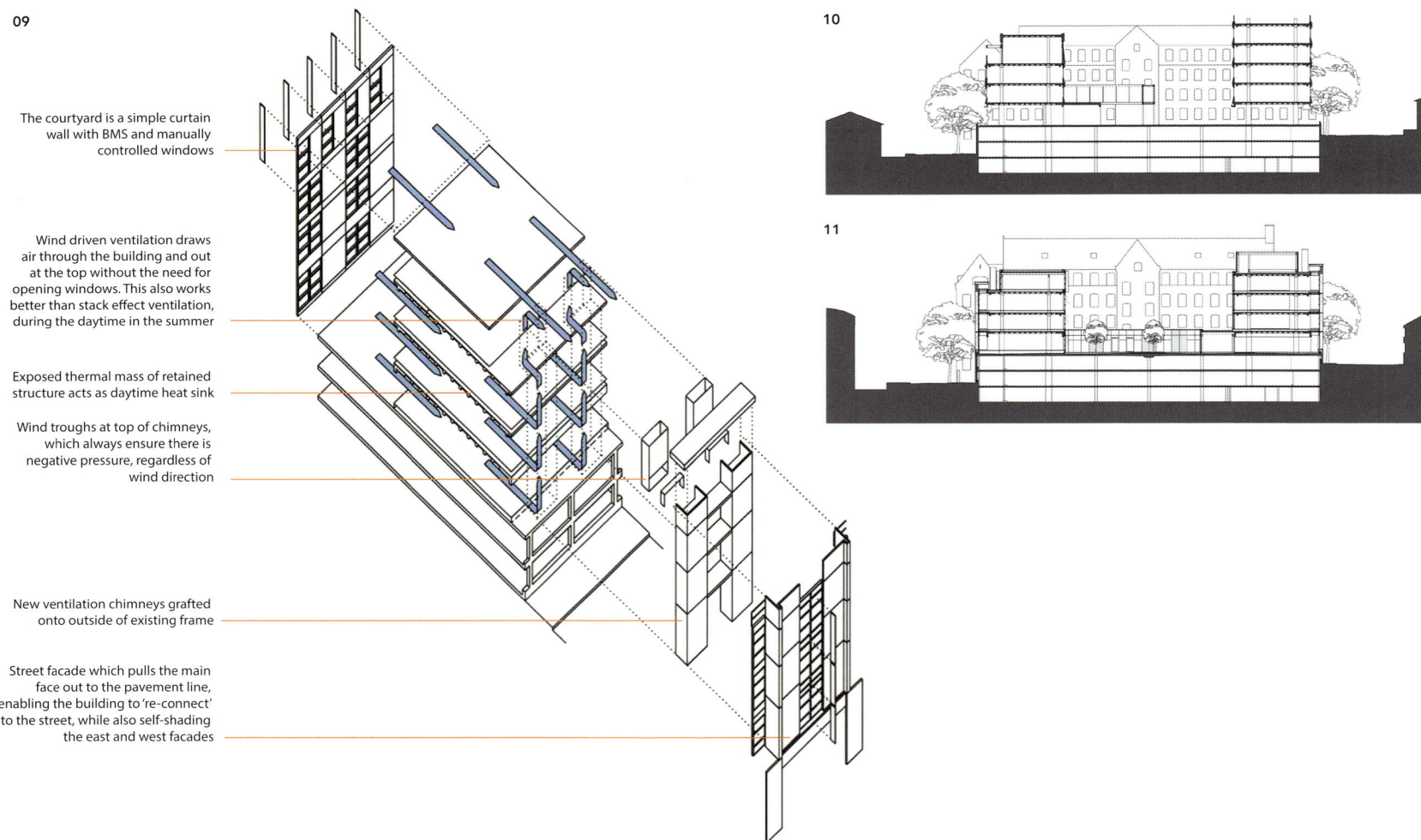

09

The courtyard is a simple curtain wall with BMS and manually controlled windows

Wind driven ventilation draws air through the building and out at the top without the need for opening windows. This also works better than stack effect ventilation, during the daytime in the summer

Exposed thermal mass of retained structure acts as daytime heat sink

Wind troughs at top of chimneys, which always ensure there is negative pressure, regardless of wind direction

New ventilation chimneys grafted onto outside of existing frame

Street facade which pulls the main face out to the pavement line, enabling the building to 're-connect' to the street, while also self-shading the east and west facades

10

11

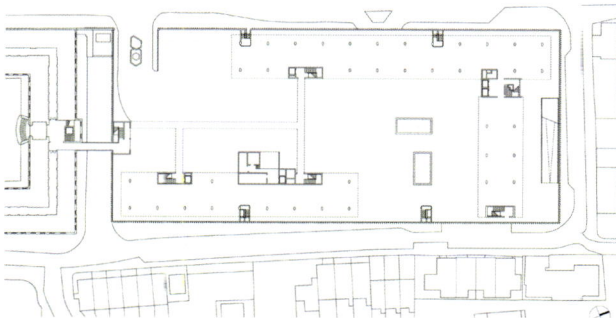

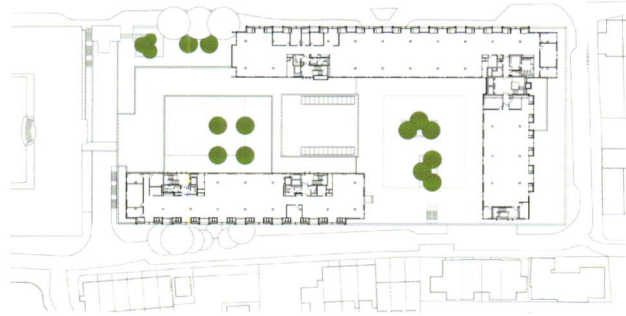

1   Entrance
2   Café
3   Lecture hall
4   Meeting rooms
5   Data centre
6   Workspace

09 The kit of parts applied to the
      retained structural frame
10 Cross-section pre-retrofit
11 Cross-section post-retrofit
12 Ground floor plan pre-retrofit
13 First floor plan post-retrofit
14 Ground floor plan post-retrofit

DATA
Energy and carbon

QUEEN ELIZABETH II COURT, WINCHESTER
Council headquarters

		Energy kWh/m²/yr	Carbon kg CO₂/m²/yr	Building information		
				Area (m²)	Occupancy (m²/person)	Hours of operation
PRE-RETROFIT	Gas	187	37.4	11,200		
Monitored	Electricity regulated	43	21.5			
	Total regulated	230	58.9			
	Electricity unregulated	43	21.5			
	Total	273	71			
POST-RETROFIT DESIGN DATA				12,591	750 workstations = 16.7 m²	09:00-17:00 + some w/e
	Predicted energy model					
	Gas	57	11		1,100 occupants = 11.4 m²	
	Electricity regulated	66	33			
	Total regulated	123	44			
	Electricity unregulated	inc.			*3:2 desk-sharing ratio	
	Renewables					
	Total	123	44	Reduced to 35 kg/m²/yr with further refinement		
ACTUAL DATA	Post-occupancy data					
	Gas	54	11			
	Electricity regulated	45	23			
	Total regulated	99	29			
	Electricity unregulated	34	17			
	Renewables	0	0			
	Total	133	46			
BENCHMARKS				Env. rating	EPC rating	DEC rating
				BREEAM 2006 Offices 'Excellent'	n/a	n/a
CONVERSION FACTORS	Gas	0.2				
kgCO₂/kWh	Electricity	0.5				

Note
1. Figures include ground floor functions (whole council)

The retrofit measures were calculated to achieve a 70% reduction in energy use. The actual energy use after two years of monitoring is indicating a 50% saving (133 down from 273 kWh/m²/yr). The difference is due to a greater intensity of use, which has led to higher than expected electricity use, and to some initial suboptimal settings on controls. The building managers are working towards identifying and tackling the energy use factors.

One problem identified was that out-of-hours use by relatively few staff requires large parts of the services system to be switched on. As a solution, the already separately zoned bookable meeting rooms are being made accessible, independently of the rest of the building, for members of staff who wish to work out of hours, so that such use requires only a small, dedicated portion of the building to be serviced.

RESOURCE USE AND WASTE
Retention of the concrete frame saved 50% of the embodied energy normally required to construct such a building, and the use of local bricks and timber-based window systems helped to significantly reduce related CO_2 emissions. A large proportion of demolition materials were recycled through the contractor's supply chain, including former precast concrete cladding panels, which were crushed off-site and reused as aggregate in other council projects.

Cost and time

The project consisted of the refurbishment, structural alteration and extension of the existing building to create 12,591 m² (gross internal floor area) of office accommodation, including Category B fit-out. The project included the removal of the cladding and strip-out of the existing building, including substantial asbestos removal As well as increasing the office area, the scheme also provided a 200-seat auditorium, a large meeting room suite, a café and restaurant, kitchen facilities and a data centre, and external works to provide two courtyards.

The total project value (including professional fees, furniture, fittings and equipment, car park repair works, limpet asbestos removal, decanting and temporary accommodation costs) was £40.166m (at a base date of November 2005).

An exercise was carried out to estimate the costs associated with providing a new-build project compared with a low-energy retrofit project. The extra over the construction costs and additional project costs, including the re-provision of car parking, was estimated as at least £30m.

An additional exercise was carried out to identify the estimated cost associated with providing a refurbished traditionally serviced office. In that case, the cost would be reduced by around £1–1.5m.

Currently, operating costs of the building are showing a £200,000 annual saving based on a number of factors, including a 50% reduction in energy consumption. This saving will increase if the 70% target is achieved.

Construction cost (per m²):

Shell and core:	£1,372
Fit-out (Categories A and B):	£365
Cladding removal and soft strip:	£114
Asbestos removal:	£90
Structural alterations:	£265
External works:	£67
Total (at November 2005):	£2,273

Project programme:

Design team appointed:	2006
Planning consent:	September 2006
Start on site:	January 2007
Practical completion:	July 2009
Occupation:	August 2009

Procurement

The design team was selected by competitive interview. The contractor was selected by two-stage tender under a 'traditional' contract:

- **first stage:** contractor selected from the local authority's framework
- **second stage:** JCT Standard Form of Building Contract Local Authorities with Quantities 1998 edition, incorporating Composite Contractor's Designed Portion and Sectional Completion Supplement 2000 edition.

A collaborative approach to financial management and control was adopted at both the pre- and post-contract stages. Using this approach at pre-contract stage generated significant benefits as a result of a number of factors. These include the team's experience of delivering similar projects; the adoption of an integrated value and risk management approach; and the integration of the main contractor to provide advice on buildability and utilisation of subcontractor expertise. These benefits, combined with the quantity surveyor's extensive cost database and component cost specialists (including envelope, data centre and engineering services), resulted in the production of robust cost plans, which in turn facilitated informed decision-making through option appraisals and sensitivity and opportunity analysis.

CREDITS

CLIENT Hampshire County Council	**ARCHITECT** Bennetts Associates
PROJECT MANAGER Mace	**M&E ENGINEER** Ernest Griffiths
MAIN CONTRACTOR BAM	**STRUCTURAL ENGINEER** Gifford
COST CONSULTANT Davis Langdon	**CFD MODELLING** EDSL
TOWN PLANNING CONSULTANT Colliers CRE	**BUILDING CONTROL** Butler and Young
ACOUSTICS Arup Acoustics	**FF&E** Hampshire County Council
	PHOTOGRAPHY Tim Crocker

Foundry Studios, London

CULLINAN STUDIO

This low-energy retrofit of an industrial building to make the architect's own canal-side office provides a fascinating case study of the trials, tribulations and ultimate success possible in refurbishment. After many years spent overcoming significant economic and planning constraints, the project achieved a BREEAM 'Excellent' rating and an energy reduction of 50% compared with the architect's previous offices in the adjacent building. The project was envisaged partly as a research exercise, to investigate the interaction between users and their building post-handover, and the new knowledge is now being exploited in the architect's professional work.

CONTEXT AND OBJECTIVES

In 1991, Edward Cullinan Architects bought a large portion of the Victorian Baldwin Terrace, comprising a three-storey 19th-century foundry and warehouse building located on the north towpath of Regent's Canal near the Angel, London. The practice moved into one end of the building and rented out the central part of the terrace as an artists' workspace and an exhibition gallery, envisaging that they would eventually retrofit it as their studio. This middle part of the terrace faces north-west onto Baldwin Terrace and south-east onto the canal, with a fine view of the

canal basin. The external walls of the existing building stand on the 36 m × 9 m site boundary with no space beyond.

The south-east wall has double-height windows and leans towards the canal. When construction began in 2011, it was 250 mm out of plumb. However, being locally listed it had to be retained, needing significant structural works to arrest its movement. The first-floor roof trusses are also locally listed and had to be retained.

The architect's associated company, The Red House, acted as the client for the project, setting the brief. The brief for the retrofit was to use two floors as its own workspace, with lettable offices on the top floor.

For the practice, trading since 2012 as Cullinan Studio, the project was about more than just creating new offices for themselves. It was also a useful research exercise, aimed at both arming them with the skills to design a retrofit project and giving them experience of being a client learning how to run a newly-completed building project.

01 Canal level double-height space along wall to canal post-retrofit
02 Canal level pre-retrofit

DESIGN: GENERAL

The first and decisive architectural moves of the retrofit were a response to two constraints of the existing foundry building: the leaning, listed south wall and its double-height windows.

Internally, steel frame shear walls articulate two circulation zones defining studio spaces. A horizontal Vierendeel truss with intermediate supports spans the shear walls to support the failing wall and creates a void inside to reveal the scale of the double-height windows. These moves stabilise the building, allow generous views and let daylight flood in. The result is a happy rearrangement of the existing building volume to connect the ground floor visually to the set-back first floor, unifying the studio despite its two levels. Keen to maximise the potential for daylight and ventilation, the windows in the listed wall were enlarged as much as permitted by conservation guidance to bring daylight to the ground and mezzanine levels.

The first floor has a particularly generous amount of natural daylight and cross-ventilation, enabled by truncating the existing north wall at first-floor cill level and inserting a continuous horizontal band of glazing in its place. Punctuated with windows, this device hangs from the new steel frame, installed to stabilise the existing building fabric and support the listed timber trusses above.

The completed building's new structural skeleton supporting the creaky original fabric defines a space that is both pleasantly bespoke and inherently of its

place. The interior is finished with simple and direct detailing, creating a calm and unpretentious setting for a design studio that takes pride in developing an architecture derived from the art of construction and the understanding of people and place.

Many iterations of the design were needed to get planning permission. The retention of the canal-side wall became a key planning condition, and a driver for the whole design. After considerable struggle, planning permission was granted in 2008, just as the recession kicked in, stalling the project for a further three years.

DESIGN: ENERGY CONSERVATION

The low-energy strategy follows a fabric-first approach, with floors, walls and roof super-insulated. Insulation in the north and south is a thick layer of recycled newspaper, up to 450 mm in places, achieving U-values of 0.1 W/m²K. The building is well sealed, with a pressure test figure of less than 5 air changes per hour – encouraging for an existing building. Having insulated to this level and provided as much natural light and natural ventilation as permissible within the constraints of the existing building, space heating and energy generation can be provided via low-energy, all-electric technologies.

The building uses an air source heat pump housed in the overrun of the old lift shaft to provide hot water and underfloor heating. Electricity is generated by 22 m² of photovoltaic (PV) panels on the south slopes of the roof, and mechanical ventilation with heat recovery services the toilets. These systems are

03 Canal wall pre-retrofit
04 Balustrade and canal wall post-retrofit
05 Street level pre-retrofit
06 Street level overlooking double-height workspace slot post-retrofit
07 Top floor pre-retrofit
08 Top floor workspace post-retrofit

controlled by a building management system (BMS) with an Internet-based dashboard to monitor energy performance and space temperatures.

The energy strategy for the project went through a number of iterations as it evolved. A plan for water-based cooling from the canal inlet could not be implemented because it was difficult to gain consent from British Waterways and the hefty fees became unaffordable. Then, although permission had been granted, the planned five wind turbines were abandoned after a wind survey revealed the low efficiency of this technology in cities, and on this site in particular. After tender, radiant ceiling panels (for space heating and cooling) and rainwater harvesting were omitted by the client on cost grounds.

USER BEHAVIOUR
AND BUILDING MANAGEMENT

The predicted energy use shows that, given the high levels of thermal insulation, most energy will be consumed for lighting and IT. Although the amount of natural daylight is limited by the existing apertures on the listed south elevation, staff are encouraged to turn off their individual task lights when possible. Passive infrared (PIR) sensors control lights in circulation spaces and toilets; to achieve the full benefit of the system, people must shut toilet doors to prevent lights turning on inadvertently as people

walk by. Computers are timed to shut off after an hour of not being used.

The architects report that the project has given them a deeper insight into the process of commissioning, interacting with and maintaining the building. In particular, the fragmented structure of the construction industry, with its separation of mechanical and electrical (M&E) subcontractors, BMS programmers and installers, complicates commissioning. Finding significant difficulties with the commissioning of the BMS and a lack of post-contract M&E, they concluded

09 Cross-section post-retrofit
10 Foundry building from Baldwin Terrace pre-retrofit
11 View along Baldwin Terrace post-retrofit

12

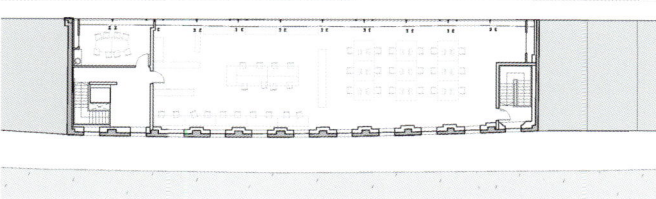

13

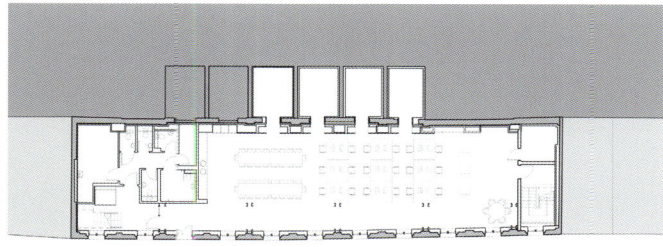

14

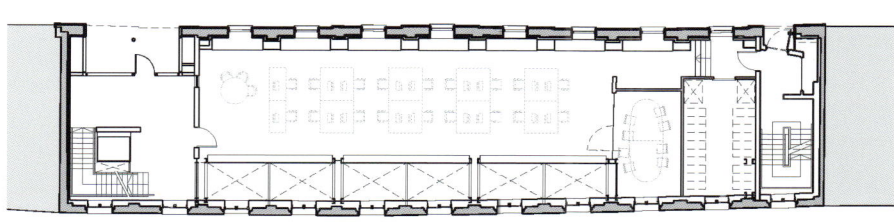

15

1 Existing fabric retained
2 Recycled newspaper
3 Natural ventilation
4 Air source heat pump
5 22 m² PV panels
6 Heat recovery
7 Underfloor heating

12 Top floor plan post-retrofit
13 Canal level plan post-retrofit
14 Street level plan post-retrofit
15 Cut-away summar sing energy
efficiency measures

that a solution, even on smaller projects, would be to ensure there is an M&E specialist within the main contractor's outfit specifically tasked with overseeing the M&E subcontractors.

Adjusting the heating is difficult because of the complexity of the BMS commissioning and control system. In contrast, natural ventilation is easily and conveniently controlled by opening or shutting windows manually. Movement sensors were a cost-reducing alternative to light sensors, but this has led to lights in circulation zones switching on in broad daylight, wasting energy. BMS metering only records energy used without separating regulated and unregulated loads, which meets the requirements of building control and BREEAM. The architects suggest that regulations should require separate metering of regulated and unregulated loads.

The architects intend to carry out a formal post-occupancy evaluation to structure feedback and capture useful knowledge for the benefit of the studio in their professional work.

DATA
Energy and carbon

FOUNDRY STUDIOS, ISLINGTON
Creative workspace

		Energy kWh/m²/yr	Carbon kg CO_2/m²/yr	Building information		
				Area (m²)	Occupancy (m²/person)	Hours of operation
PRE-RETROFIT (old office)	Gas	110.00	22.00			
Estimated	Electricity regulated	n/a				
	Total regulated	n/a				
	Electricity	111.00	55.50			
	Total	221.00	22.00			
POST-RETROFIT DESIGN DATA				858	12	Office hours
	Part L model (2010)		12.80 BER			
	Predicted energy model					
	Gas	0.00	0.00			
	Electricity lighting	26.21	13.11			
	Total regulated	26.21	13.11			
	Electricity unregulated	37.45	18.73			
	Renewables (inc. above)	-3.26	-1.72			
	Total	63.66	31.83			
ACTUAL DATA	Post-occupancy data					
	Gas	n/a	n/a			
	Electricity regulated	n/a				
	Total regulated	n/a				
	Electricity total	115.00	57.50			
	Renewables	n/a				
	Total	115.00	57.50			

BENCHMARKS		Env. rating	EPC rating
		BREEAM 2008 'Excellent'	B (28)

CONVERSION FACTORS	Gas	0.2
kgCO_2/kWh	Electricity	0.5

Note
1. CO_2 by cubic volume:
   Building volume 2,651m³
   CO_2 emissions (calculated) 10.30kgCO_2/m³/yr
   CO_2 emissions (actual) 18.61 kgCO_2/m³/yr

Energy consumption in the first year of use was almost double the level predicted; work to establish the cause is ongoing with the M&E subcontractors during the defects rectification period. There is a problem with the heat pump controls, the resolution of which will reduce electricity consumption. Energy saving over the previous office was modelled at 75%, but the current saving is 50%. The PV panels make only a modest contribution; it would have been better to install a larger array, which calculations show would have made the building self-sufficient in energy. The cost of such an array would have been about the same as the VAT paid on the retrofit project. If, as many have argued, refurbishment was zero (or minimum) rated for VAT it would have incentivised a wiser long-term investment here.

Resources and waste

As an exercise in what could be done on a limited budget, Cullinan Studio focused on reducing embodied energy. To that end they reused as much of the existing structure as possible. As a result, only one-third of the materials in the completed project were new. Construction waste was monitored to BREEAM requirements. The practice continually monitors and records office waste; they chose a provider who has good recycling capacity and provides detailed breakdowns on recycling quantities, which are used to inform the carbon footprint. There is a plan to compost food waste in a small planting area outside the office.

Cost and time

Gross internal area:	785 m²
Cost (per m²):	£1,720 net, excluding contractor OH&P
Contract duration:	11 months
Completion:	September 2012

16

16 The new entrance

Procurement

The building was built using the JCT Design and Build Contract (2005 edition).

CREDITS

CLIENT
The Red House

CONTRACTOR
Jerram Falkus

M&E ENGINEER
Max Fordham LLP

ARCHITECT
Cullinan Studio

STRUCTURAL ENGINEER
Cullinan Studio

PHOTOGRAPHY
Tim Soar and Cullinan Studio

Somerset House East Wing, London

KING'S COLLEGE LONDON

One of the first instances of a refurbishment and adaptation of a Grade I listed building to be designed to achieve a BREEAM 'Excellent' rating, this project integrates the East Wing of Somerset House with King's College London's Strand Campus next door, making new connections across the site. This project forms the third phase of a long-term masterplan, with wide-ranging aims: the creation of a modern higher education environment that is attractive to students and staff from all over the world, with universal access, architectural excellence and high environmental ambitions. It shows what can be achieved by a confident and expert client, working in close co-operation with the architect, engineers and other specialists.

01 View across the Great Court, Somerset House
02 Front 'area' before retrofit showing bridge partially removed and entrance blocked with railings. The new design reinstated the bridge and reactivated the entrance

CONTEXT AND OBJECTIVES

Somerset House, designed by Sir William Chambers in 1776 and completed over the following 25 years or so, has been used by a number of different government offices over its lifetime and today is home to a variety of cultural and artistic organisations. It is well known and loved in London as a major public attraction, accommodating several important art collections, while its Grand Court, now famous for its fountains, is the site of a popular temporary ice rink at Christmastime.

Secured on a 78-year lease in 2009 from the Somerset House Trust, the refurbished East Wing, originally completed around 1801, is part of a campus masterplan that also includes a number of listed Georgian, Victorian and Edwardian buildings and the old Strand underground station, along with some 1970s brutalist architecture. After the 18-month refurbishment, completed in 2012, it forms the western boundary of King's College London's Strand Campus.

Tenant and landlord had a mutual interest in promoting cross-fertilisation between their two organisations, seeing the building's potential to enrich each other's ultimate cultural, artistic and educational purposes. The brief specified that the spaces should encourage this interaction, requiring significant improvements to the accessibility, accommodation and support facilities.

Somerset House was originally designed as a series of 'town houses' containing government offices, each with its own entrance, vertical circulation and associated physical separation. Subsequent use by HM Custom and Excise had resulted in the removal of much of this historic subdivision and the loss of many of the original cantilevered staircases. Multiple further alterations left a building lacking any sense of internal clarity, with poor and inefficient use of space and in critical need of significant restoration and de-cluttering.

Key elements of the brief were to optimise the quality of the working environments – to encourage staff productivity – and to rethink the internal circulation routes. Overarching a number of functional and aesthetic aims, the brief required reductions in the building's energy and carbon footprints. As a publicly committed advocate of environmental sustainability across its entire estate, King's required that the building's environmental performance should be optimised in balance with conservation and user needs.

The result is that new galleries and exhibition facilities, together with café and toilets, now occupy the entrance level, accessed from new links from King's campus, the Quad and the Great Court. Predominantly used by Somerset House Trust, this accommodation is adjacent to King's Executive Centre, which comprises new meeting, seminar and exhibition spaces at first and lower ground floor levels. The arrangement deliberately forces the activities of King's and the Trust to overlap, recognising the virtue in cultural exchange that arises from chance encounters. The upper levels of the building are occupied by The Dickson Poon School of Law.

DESIGN: GENERAL

Having worked with King's before, the architects BDP had a ready understanding of the client's vision and objectives. Their adaptive design, generated in close co-operation with the client and supported by heritage specialists, was primarily conservation-led, with time taken to ensure that energy improvement measures would be acceptable from a heritage and building conservation point of view.

The design approach incorporated a radical reordering of the internal organisation to improve spatial clarity, physical connections and environmental

efficiency. It followed the principle of revealing the best of what was already there but obscured and overlaid by later additions. A striking example of this is the extension of the central cantilevered stone stair. The missing elements of the staircase were rebuilt using traditional techniques, returning it to its original height and proportions after almost a century. The primary changes include the rationalisation of horizontal connections to improve physical and visual links along the length of, and up through, the building. New entrances into the building, together with new stair and lift access between the various levels, enhance connectivity and access to all levels of

03 + 04 Light wells opened up along the main axis of the building are part of the new ventilation strategy and establish visual links
05 + 06 Some areas have simply been restored and provided with new lighting

FIRE EXIT

TO LIFT

the building. The design anticipates a possible future 'front door' to King's from the Great Court.

The original spatial qualities of many of the interiors have been re-established and a number of historic elements have been integrated to create a coherent whole. Features such as cornices and fireplaces have been restored, while the interior decoration is fresh and simple, complementing the Georgian style and colour palette. Following the discovery of ancient foundations below one of the basement rooms, a breakout space with a glass floor has been created.

Public and stakeholder consultations were carried out between October and December 2009, culminating in a public exhibition of the design proposals, held in the East Wing spaces.

There was extensive statutory consultation with Westminster City Council and English Heritage. Their involvement in the project continued well into the construction stage, as discoveries (particularly in the basement, where Tudor and Saxon foundations were uncovered during the demolition works) required further discussions and agreement on how to address the significance of some of the remains. The Georgian Group also gave advice.

07 New thermally high-performing roof lanterns along the main circulation axis bring more light into lower floors, assist the ventilation strategy and reduce energy leakage

08 + 09 Removing concrete bridges introduced by the previous occupiers has reinstated the spatial clarity of the original stone staircases

07

08

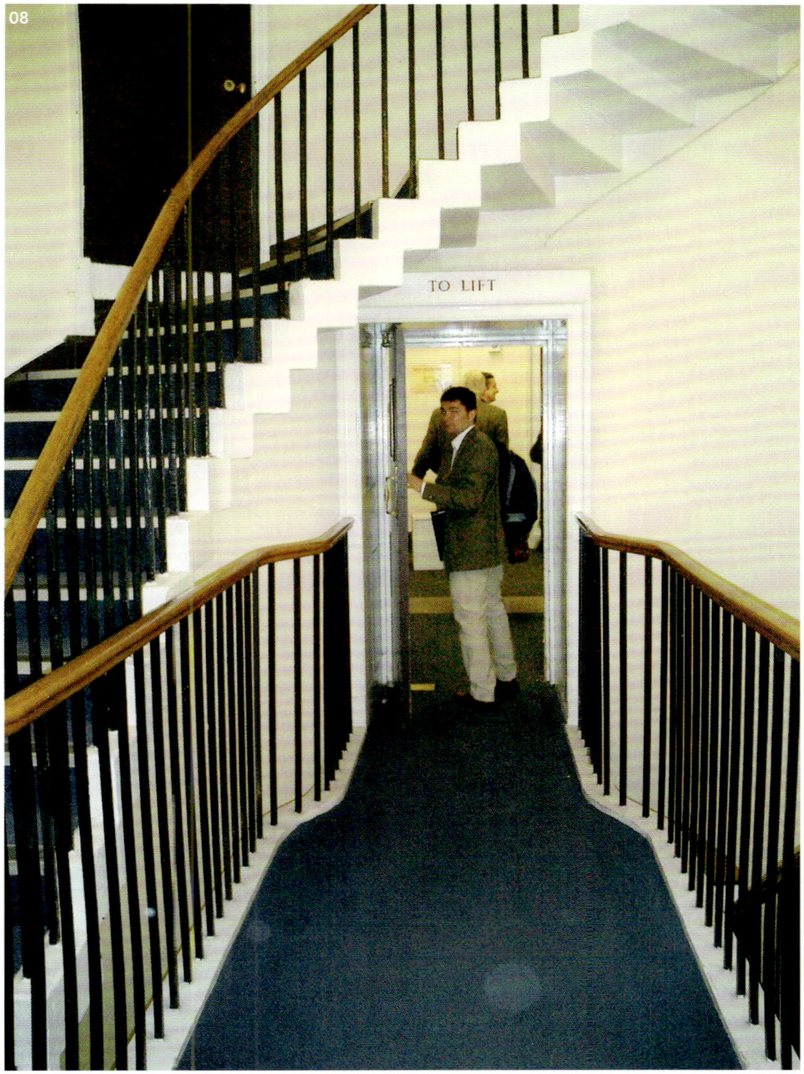

09

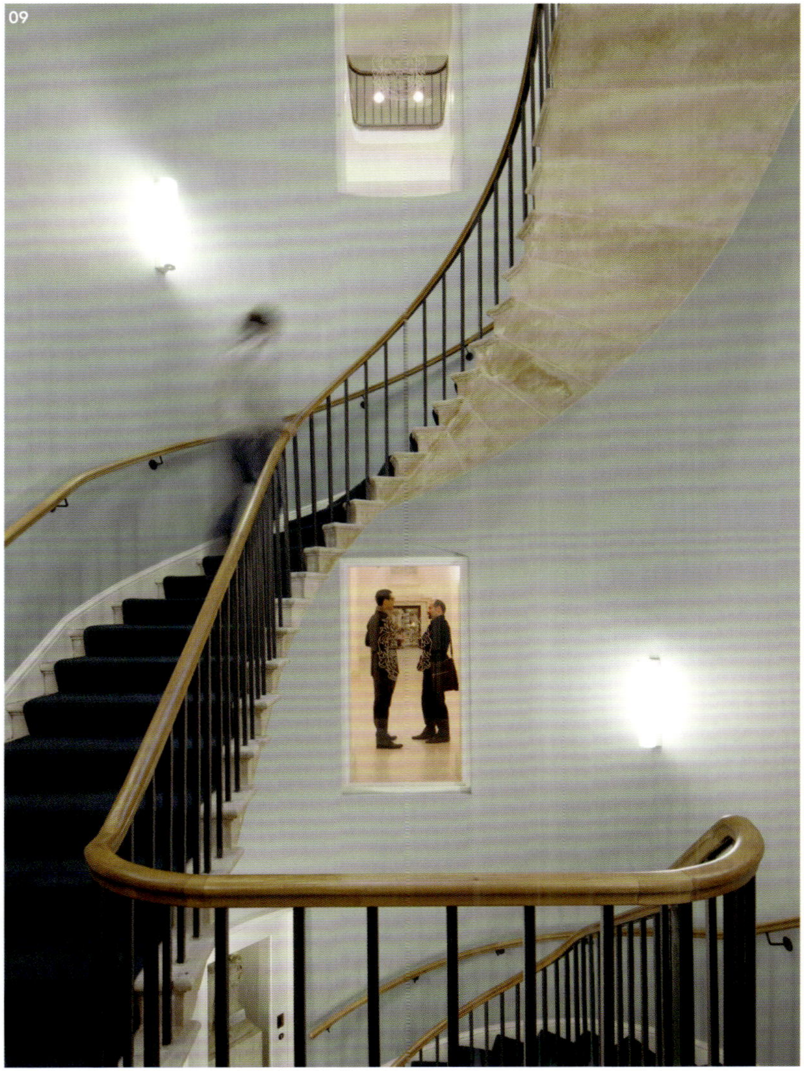

DESIGN: ENERGY CONSERVATION

Considerable attention was given to determining comfort criteria, to assessing technical feasibility and whole-life costs, and to ensuring that energy use and carbon emissions were considered from the outset. A detailed thermal model, used to predict CO_2 emissions and thermal comfort under a variety of design scenarios, was adopted as a critical design tool. The design team also used non-intrusive thermography surveys to identify air leaks and confirm the benefits of improved airtightness. The resulting energy-saving strategy combines reduced demand, improved fabric and efficient building services.

The Grade I listing ruled out the standard passive tactics, such as external or internal wall insulation, replacing windows or installing secondary glazing. However, the building has a range of inherent features that, with the right strategies, help to keep energy consumption down without clashing with the historic fabric or detracting from its architectural quality. These include generous floor-to-ceiling heights on most floors, important for the ventilation strategy, thick external walls, providing thermal mass and some insulation, and glazing that admits daylight without unmanageable summer overheating from the sun.

Where these features were missing, the space planning helped to compensate. For instance, the naturally dark lower basement levels were the obvious choice for seminar rooms and lecture theatres, as it is often necessary to exclude daylight during their use. The design team's selection of passive measures included high-performance insulation below the ground floor slab and, in the roof, low embodied energy insulation manufactured from waste wood fibre products, with environmental and conservation benefits. Following experiments to test its visual impact, removable film was added to all windows to help to control solar gain. New window seals were fitted to reduce air leakage where most required, and hidden roof valleys were given new double-glazed windows and external louvres.

The top floor posed a particular design challenge, being exposed to unwanted heat gains from the slated roof and lacking the tall floor-to-ceiling heights of the other storeys. The designers' response was to embed a phase-change material (PCM) into the roof plasterboard finish. This elegant solution provides thermal storage that is not reliant on mass and minimises the risk of overheating, all without impinging on the internal proportions. A typical PCM product with a thickness of 5 mm is equivalent in thermal mass to 60–80 mm of concrete. Modelling predicted that PCM would reduce periods when temperatures are above 26°C by approximately 15%, and would reduce peak temperatures by approximately 3°C.

10 + 11 Previously dark spaces in the basement have been opened up as part of the new circulation and lighting strategy

12 + 13 A new bronze and glass bridge spans across the east light well, facing King's quadrangle, for access to its ground floor spaces

14 Saxon and Tudor foundations were discovered during the construction works under one of the basement rooms; now on display under a glass floor

There is mixed-mode ventilation throughout, using opening windows wherever and whenever possible. This is supplemented by mechanical ventilation, which neatly uses existing fireplace flues for drawing in and distributing fresh air, a strategy that necessitated the careful mapping of existing chimney flues and reopening of any that had become blocked or compromised.

During favourable external conditions all accommodation is naturally ventilated through existing sash windows, with enhanced ventilation across rooms and up through the building via staircases and voids along the central circulation spine. During unfavourable external conditions (acoustics, temperature, humidity and wind) windows are closed; this activates the mechanical ventilation system. Tempered external air is distributed, via small-scale plant rooms at upper floor level, through liners inserted within chimney flues and discharged via fireplaces into rooms, at low velocity. This strategy limits intrusive ductwork. Microswitches ensure that when windows are opened, the mechanical system closes down to conserve energy.

To facilitate cross-ventilation, new attenuated openings have been formed between rooms and the central spine corridor, utilising the precedent of overlights above internal doors. This maximises the efficiency of both the natural and mechanical ventilation strategies, while also satisfying the acoustic brief. Warmed air circulating up and through the central voids is extracted at roof level.

The M&E services are connected to central plant and are ready to be connected to a proposed future combined heat and power (CHP) supply. Sub-meters are deployed extensively, ready to be linked to King's building management system (BMS).

Finally, there are new energy-efficient lighting, appliances and controls throughout, along with efficient water appliances and a leak-detection system, also linked to the BMS.

USER BEHAVIOUR AND BUILDING MANAGEMENT

The project team was conscious of the importance of occupant behaviour for the ultimate success of the energy conservation effort. Future building occupiers were able to keep up with progress through an internal website, with plans and updates. As part of this process, alumni and future building occupiers were invited on site visits during the construction stage. Upon moving in, occupants received a short user guide and user manuals for various features, and were offered training. The project also invited community involvement by maintaining a website to communicate the sustainability strategy and project progress.

In accordance with its energy and environmental policies, King's will be monitoring and controlling the building's performance.

15

15 Cross-section post-retrofit
16 Ground floor plan post-retrofit

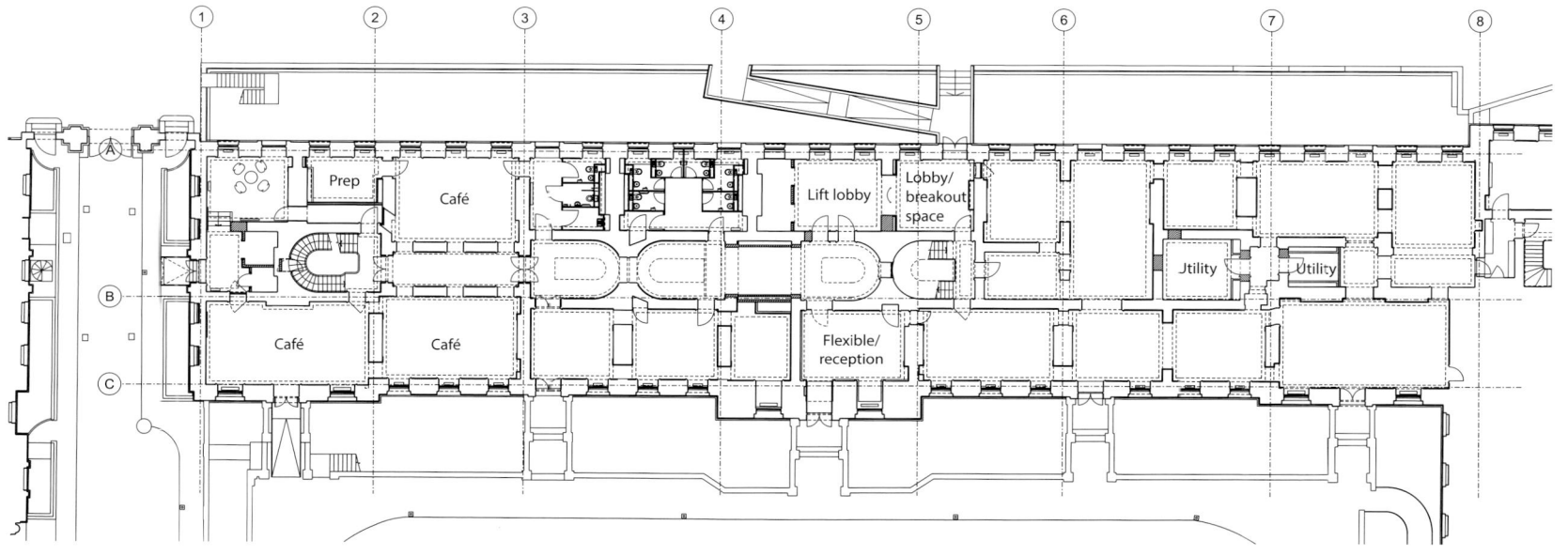

Prep

Café

Lift lobby

Lobby/
breakout
space

Café

Café

Utility

Utility

Flexible/
reception

DATA

Energy and carbon

SOMERSET HOUSE EAST WING, LONDON
University accommodation

		Energy kWh/m²/yr	Carbon kg CO$_2$/m²/yr	Area (m²)	Building information Occupancy (m²/person)	Hours of operation
PRE-RETROFIT	Gas			7,100		
Monitored	Electricity regulated					
	Total regulated					
	Electricity unregulated					
	Total	n/a	n/a			
POST-RETROFIT DESIGN DATA				6,500		
	Part L model (2006)	see Note 1	23.1			
	Predicted energy model					
	Gas	66.9	13.38			
	Electricity regulated	19.2	9.60			
	Total regulated	86.1	22.98			
	Electricity unregulated	38.5	19.25			
	Electricity regulated and non-regulated	57.7	28.85			
	Renewables		−12.35	CHP see Note 2		
	Total		29.88			
ACTUAL DATA	Post-occupancy data					
	Gas					
	Electricity regulated					
	Total regulated					
	Electricity non-regulated					
	Renewables					
	Total	n/a	n/a			
BENCHMARKS				Env. rating	EPC rating	DEC rating
				BREEAM 2008 'Excellent'	B	
CONVERSION FACTORS	Gas	0.2				
kgCO$_2$/kWh	Electricity	0.5				

Notes:
1. Target CO$_2$ emission rate (TER) = 31.3 kgCO$_2$/m²/yr – at RIBA Work Stage D. Occupancy profiles changed since.
2. With CHP the emissions are approximately 30 kgCO$_2$/m²/yr depending on equipment selection; without CHP the value is 42.3 kgCO$_2$/m²/yr

Modelling indicates that energy performance will be comparable with that of the best newly built office accommodation. Possible reasons for such a performance level include the following:

- Regulated consumption in office buildings is dominated by lighting, fans and cooling. Here, lighting was designed to be very efficient, and only some parts of the building are mechanically ventilated and cooled.

- The good performance is in line with modelling of comparable naturally ventilated or mixed-mode buildings. The modelling method favours natural ventilation, and probably incorporates optimistic assumptions in terms of occupant behaviour: for example, the air change rates are set at Building Regulations Part F background ventilation rates, while in practice windows may be open more often, which would drive up space heating demands.

- To some extent the modelled predictions reflect the inherent qualities of the building as an office: while insulation levels and glazing U-values are not very good, there is high thermal mass and glazing proportions are relatively low. In any case, retaining heat with high insulation levels tends to be counterproductive for modelled results for office buildings.

Of course, a model can only predict, and so the real test will be the actual in-use figures, which are not available at the time of writing.

When CHP plant is factored in, the model indicates higher savings than would be expected in a new office building. This is because the space heating load is relatively high and CHP is particularly efficient in contributing to space heating. The contribution of CHP is a 'potential' one as it is not yet installed – a decision on the planned local heating network is still awaited.

Resources and waste

As far as possible, the existing fabric was kept. New elements were fabricated and installed using mainly recycled materials and traditional techniques. Wastage of material was controlled, and as much as possible waste was either reused or recycled by the contractor.

Cost and time

Gross internal area: 6,500 m² (note: this includes the whole of the ground floor, although only 150 m² was fully refurbished for King's College London; the rest was finished as shell and core, for Somerset House Trust to finish).

Construction cost:	£16.7m
Cost per m²:	£2,569
Contract commencement:	August 2010
Completion:	February 2012
Contract duration:	18 months

Procurement

The main contractor was appointed in August 2010 through competitive tender. Five companies were shortlisted initially and were interviewed in May 2010.

The project used a traditional contract (JCT Standard Building Contract Without Quantities 2005 edition [Revision 2 2009]), and was fully designed by the design team.

CREDITS

CLIENT AND TENANT
King's College London

PROJECT MANAGER
Gardiner & Theobald Management Services

MAIN CONTRACTOR
Wates Construction Ltd

QUANTITY SURVEYOR
Turner & Townsend

LANDLORD
Somerset House Trust

ACOUSTIC CONSULTANT
Sandy Brown Associates

PLANNING CONSULTANT
Gerald Eve LLP

HERITAGE CONSULTANT
Alan Baxter & Associates

PUBLIC CONSULTATION
Four Communications

ARCHITECT AND CONSERVATION SPECIALIST
Building Design Partnership

M&E ENGINEER
Hoare Lea

C&S ENGINEER
Ramboll UK

DDA CONSULTANT
David Bonnett Associates

FIRE CONSULTANT
Zeta Services Fire Risk Management

APPROVED INSPECTOR
Dunwoody Building Legislation

CDM COORDINATOR
RAB Consulting

LIFT CONSULTANT
JB Lift Consultants

PHOTOGRAPHY
Sanna Fisher-Payne and David Barbour

Stable Block, Morden Hall Park, Surrey

THE NATIONAL TRUST

Historic buildings often present limited opportunities for achieving high levels of energy efficiency in the fabric. In this demonstration project, while pragmatic steps were taken to significantly improve the thermal performance of the envelope, the main target was to reduce reliance on fossil fuels and eliminate carbon emissions through on-site energy generation and heat transfer. It shows what is possible in unpromising contexts if the client is determined to create an exemplar and the design team is committed to exploring a range of technologies.

CONTEXT AND OBJECTIVES

Morden Hall Park has a long history, dating back to the late 18th century. Most of the 50-hectare estate was bequeathed to the National Trust in 1941, including a snuff mill, powered by the River Wandle, which flows through the park, and the adjacent stable block.

In 2008, nine municipal authorities in five north-west European countries set up the Living Green Project (www.livinggreen.eu), part funded by the European Union, to renovate historic urban buildings and to 'promote knowledge on sustainable renovation'. The National Trust joined this project with the

refurbishment of the stable block, with the aim of creating an exemplar of historic building reuse. The Heritage Lottery fund also contributed 40% of the costs to the project.

The stable yard, built in 1879, though not listed on the English Heritage register, is locally listed as a 'non statutory building important to the local community'. The stable block is arranged around a courtyard as a C-shaped single-storey block, with the fourth side, the entrance side, completed by a gated wall.

The aim of the retrofit project was to 'achieve a carbon neutral building, generate 100% of energy on site, respect cultural and architectural values, comply with the demands generated by climate change and demonstrate these techniques to the visiting public'.

There was a strong didactic intention behind the project. During construction the public were invited to learn about the construction process through a number of taster days and open events themed with relevant topics, such as insulation materials, renewable technologies, lime plaster and waterwheels.

01 Exhibition and learning space, west wing of the stables post-retrofit

02 Same space pre-retrofit

DESIGN: GENERAL

The retrofitted stable block houses a 'Living Green' exhibition hall, a cafeteria, National Trust offices, a second-hand bookshop and new WC facilities.

The stable yard had been used as a maintenance depot for the park in recent years with little alteration from its original purpose until the 1940s and barely altered since. Its brick walls laid in lime mortar are 450 mm thick at the base and then thin to 215 mm in panels between external piers.

The basis of the environmental strategy is an internal layout that seeks to optimise the benefits of the building's orientation. The offices, for example, are in the west wing, so they enjoy solar gain in the morning but are in shadow in the afternoon, preventing overheating. The design works very much with the grain of the unpretentious and simple details of the original building, making the most of the spaces which, despite their simple form, have considerable internal variety.

The main intervention into the existing construction is the extensive internal insulation of walls and roof; internal so as to preserve the external appearance, a common theme in such projects. Natural insulation materials, typically 240 mm thick and of a number of differed kinds to demonstrate alternative options, were used together with lime plasters to allow the structure to breathe. The hemp insulation used in the roof and the Edenbloc insulation for walls, made from recycled wool, were both UK produced. Cork insulation was imported by ship from Portugal. Polyisocyanurate and polyurethane insulation was avoided on the grounds of

its impermeability and environmental impact, despite its higher insulation value. Glazed 'cut-outs' reveal the wall construction as part of the educational agenda of the Living Green project.

The floor has been dug out, insulation and underfloor heating installed and the original stable bricks and decorative tiles reinstated. The large windows of the main exhibition spaces are triple glazed. Original windows have been retained throughout and their thermal performance enhanced with either double-glazed tilt-and-turn windows added on the inside, or thin double-glazing with warm spacer technology and Pilkington Spacia vacuum glazing installed within the original window rebates. The occupied areas of the building are naturally ventilated via the conserved

original roof turrets and opening windows. There is no mechanical cooling.

Thermal imaging was used upon completion of the envelope to verify the continuity of the insulation and intelligent vapour membrane, with no air leakages or cold bridging found.

03 Stable block post-retrofit
04 Courtyard pre-retrofit
05 Courtyard post-retrofit, showing the exhibition wing (right) and PV panels and café (left)
06 Horse stalls pre-retrofit
07 Horse stalls converted for use as book and market stalls. The original floor has been reinstated over insulation

The approach to the thermal performance improvements was influenced by the Passivhaus methodology, although the Passive House Planning Package was not used. In addition to the high levels of insulation, airtightness was also addressed – an air permeability of 8.1 m³/(h×m²) was achieved, which is good for an old building of this kind.

08

DESIGN: ENERGY CONSERVATION

Having reduced energy demand by upgrading the thermal performance of the building through relatively simple but effective techniques, the design team set out to eliminate net energy import through low-carbon on-site energy generation, deliberately showcasing technologies that householders could install in their own homes. Photovoltaic (PV) panels, hybrid PV/solar thermal panels and 'solar slates' on the roofs generate an estimated 6,235 kWh of electricity and 12,570 kWh of heat for water per year. An air-source heat pump and a wood-fired boiler contribute another 18,000 kWh for space heating.

The fixing of the solar panels was detailed so they nestle snugly into the profile of the old building. The solar slates, of course, marry into the existing surface directly, although their output is relatively low.

Additional elements of the retrofit that contribute to reducing the environmental impact of the building include:

- Lithotherm clay plates for the underfloor heating, made from recycled clay products
- recycled glass worktops, made from bottles from the National Trust cafeteria
- a green roof to the cycle shelter
- rainwater harvesting for the WCs

- innovative sanitaryware that incorporates an integral wash-hand basin within the WC cistern, providing waste water from handwashing straight into the WC cistern for flushing
- waterless urinals and low-flow taps
- the use of thermal storage to make the best use of the mixed renewable and low-carbon heat sources: high-capacity hot water storage absorbs heat from the solar thermal panels during daylight hours, which is then slowly released at night, smoothing out peaks and troughs in the 24-hour cycle
- simple, intuitive user controls
- detailed sub-metering for energy and water use analysis and for public display in an easy to understand format
- passive presence and absence detectors, which control interior lighting, ensuring lighting is only on when required
- environmentally friendly paints, including water-based paint for external joinery and casein-bound distemper for internal walls.

The project harnesses the power of the River Wandle via an electricity-generating Archimedean screw turbine at the weir head, estimated to generate 60,000 kWh per year. This electricity feeds the stable yard, with the excess feeding into the National Grid.

08 Archimedean screw turbine
09 Exhibition space cork insulation being installed
10 Exhibition space interior after strip-out
11 Vapour and airtightness membrane
12 Exhibition and learning space, looking out to the courtyard

This excess mainly occurs in winter, when the river is in full flow. In summer, the water flow is much reduced, although it is constantly fed by the outflow of treated water from a sewage treatment plant upstream. The one remaining historic waterwheel has been conserved and some facsimile paddles reinstated, marking the generation of mechanical power that sustained Morden Hall's industrial past.

USER BEHAVIOUR AND BUILDING MANAGEMENT

Compared with workplaces or formal education buildings, user behaviour in a visitor centre such as this is less amenable to direction. However, the aim of the project is to educate people in, and raise awareness of, energy conservation and sustainability, with an eventual impact on the way people use buildings. The energy-related plant has been placed behind a glass screen and is therefore highly visible. Familiarisation of the centre manager with the controls is ongoing.

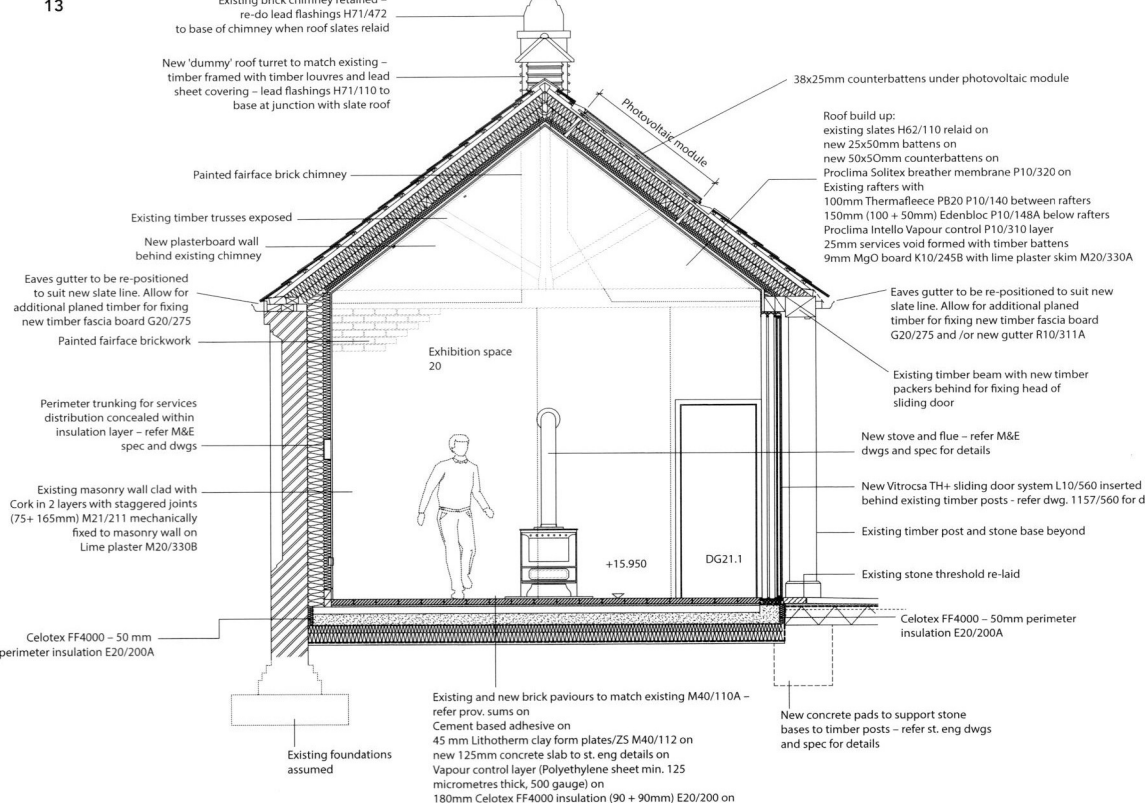

13

Existing brick chimney retained – re-do lead flashings H71/472 to base of chimney when roof slates relaid

New 'dummy' roof turret to match existing – timber framed with timber louvres and lead sheet covering – lead flashings H71/110 to base at junction with slate roof

Painted fairface brick chimney

Existing timber trusses exposed

New plasterboard wall behind existing chimney

Eaves gutter to be re-positioned to suit new slate line. Allow for additional planed timber for fixing new timber fascia board G20/275

Painted fairface brickwork

Perimeter trunking for services distribution concealed within insulation layer – refer M&E spec and dwgs

Existing masonry wall clad with Cork in 2 layers with staggered joints (75+ 165mm) M21/211 mechanically fixed to masonry wall on Lime plaster M20/330B

Celotex FF4000 – 50 mm perimeter insulation E20/200A

Existing foundations assumed

38x25mm counterbattens under photovoltaic module

Photovoltaic module

Roof build up:
existing slates H62/110 relaid on
new 25x50mm battens on
new 50x50mm counterbattens on
Proclima Solitex breather membrane P10/320 on
Existing rafters with
100mm Thermafleece PB20 P10/140 between rafters
150mm (100 + 50mm) Edenbloc P10/148A below rafters
Proclima Intello Vapour control P10/310 layer
25mm services void formed with timber battens
9mm MgO board K10/245B with lime plaster skim M20/330A

Eaves gutter to be re-positioned to suit new slate line. Allow for additional planed timber for fixing new timber fascia board G20/275 and /or new gutter R10/311A

Existing timber beam with new timber packers behind for fixing head of sliding door

New stove and flue – refer M&E dwgs and spec for details

New Vitrocsa TH+ sliding door system L10/560 inserted behind existing timber posts - refer dwg. 1157/560 for details

Existing timber post and stone base beyond

Existing stone threshold re-laid

Celotex FF4000 – 50mm perimeter insulation E20/200A

Exhibition space 20

+15.950

DG21.1

Existing and new brick paviours to match existing M40/110A – refer prov. sums on
Cement based adhesive on
45 mm Lithotherm clay form plates/ZS M40/112 on
new 125mm concrete slab to st. eng details on
Vapour control layer (Polyethylene sheet min. 125 micrometres thick, 500 gauge) on
180mm Celotex FF4000 insulation (90 + 90mm) E20/200 on
Visqueen EcoMembrane DPM 2000 gauge J40/120B on
25mm sand blinding

New concrete pads to support stone bases to timber posts – refer st. eng dwgs and spec for details

14

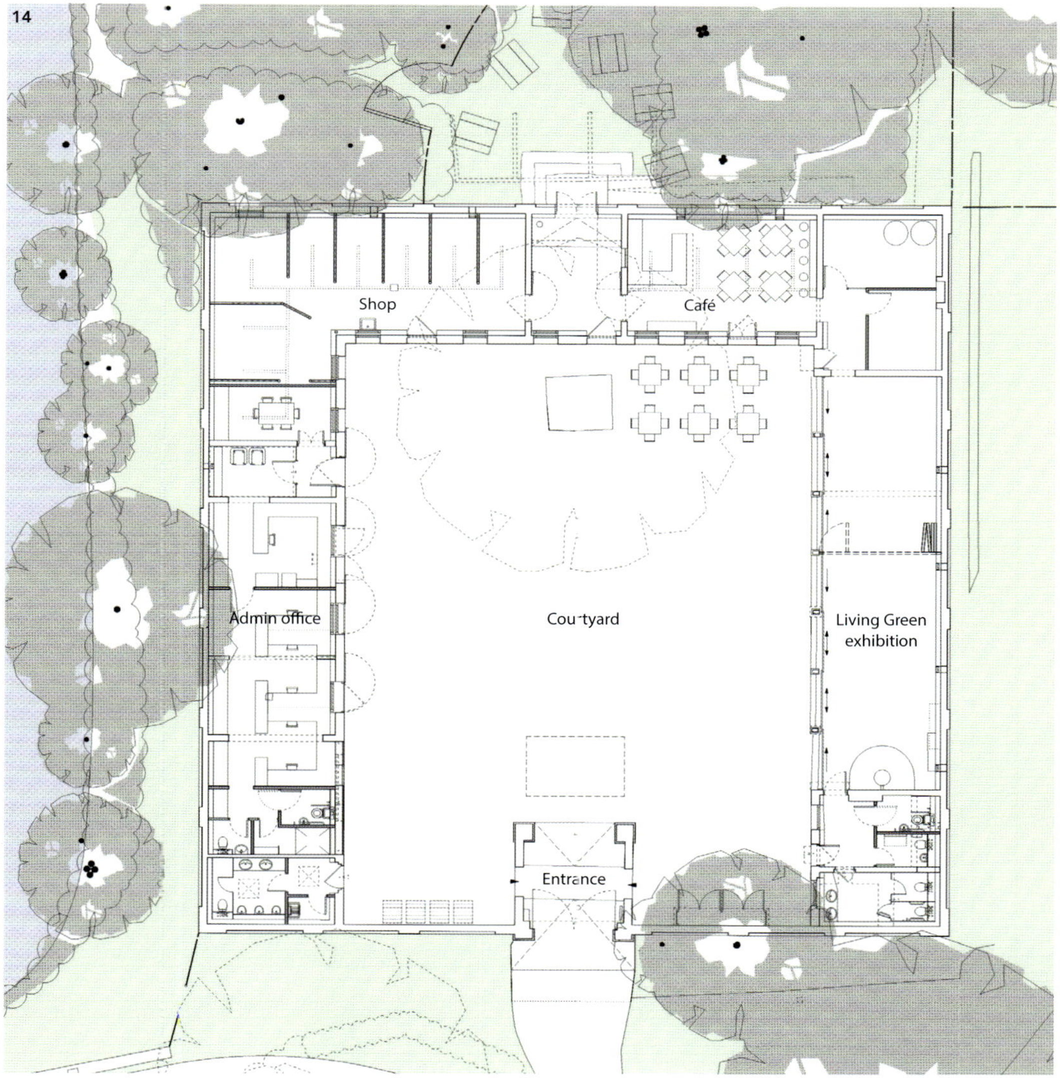

Shop

Café

Admin office

Cou·tyard

Living Green
exhibition

Entrance

15

16

DATA

Energy and carbon

MORDEN HALL PARK, STABLE BLOCK
Visitor centre

		Energy kWh/m²/yr	Carbon kg CO₂ / m²/yr	Building information		
				Area (m²)	Occupancy (m²/person)	Hours of operation
PRE-RETROFIT	Gas			468.00 gross		
Monitored	Electricity regulated			448.00 treated		
	Total regulated					
	Electricity unregulated					
	Total	n/a	n/a			
POST-RETROFIT DESIGN DATA						
	Predicted energy model					
	Space heating & hot water	71.89	14.38			
	Electricity regulated	74.35	37.18			
	Total regulated	146.24	51.55			
	Electricity unregulated	61.34	30.67			
	Total energy demand	207.58	82.22			
	Renewable heat	−68.37	−13.67			
	Renewable Power	−147.85	−73.92	See breakdown below		
	Total Renewables	−216.22	−87.60	See breakdown below		
	Net energy/CO₂	−8.64	−5.38			
ACTUAL DATA	Post-occupancy data	n/a	n/a			
	Gas					
	Electricity regulated					
	Biomass					
	Total regulated					
	Electricity unregulated					
	Renewables					
	Total	n/a	n/a			
BENCHMARKS	ECON 73 Typical Practice Primary School	201.00		Env. rating	EPC rating	
	ECON 73 Good Practice Primary School	146.00		n/a		
CONVERSION FACTORS	Gas	0.2				
kgCO₂/kWh	Electricity	0.5				

RENEWABLES	Hybrid PVT	BE PV	Solar slate	Woodburner	Turbine	ASHP	Total
Power kWh/yr	3,400	1,510	1,325		60,000		66,235
Heat kWh/yr	12,570			8,200		9,820	30,590

Notes

1. PVT: Photovoltaic/solar thermal; BE PV: Building integrated PVs; ASHP: Air Source Heat Pump
2. The turbine is an Archimedean screw powered by the River Wandle

In the first year the building used 52 MWh of electricity, compared with the estimated 56 MWh. If the building had been renovated to 2009 Building Regulations standards the estimated energy use would have been 127 MWh, more than double.

The electricity generated by the turbine makes this project a net exporter of energy to the National Grid.

Resources and waste

The National Trust has an extensive recycling programme.

Cost and time

Gross internal area:	468 m²
Net internal area:	448 m²
Construction cost:	£1.21m
Cost per m²:	£2,591
Contract duration:	12 months
Completion:	November 2011
	(turbine: November 2012)

Procurement

The project was procured using a traditional form of contract (JCT Standard Building Contract with Quantities) in two separate contracts: enabling works and main contract.

CREDITS

CLIENT
The National Trust

CONTRACTOR
R Durtnell & Sons Ltd

COST CONSULTANT
Davis Langdon

MILLS ADVISER
Purcell Miller Tritton

ARCHITECT
Cowper Griffiths

M&E ENGINEER
Ridge & Partners

STRUCTURAL ENGINEER
Crofton Design

PHOTOGRAPHY
Louis Sinclair

17 Café post-retrofit

Moray Council Headquarters, Elgin

MORAY COUNCIL

By creating a stimulating and productive workplace out of a disused supermarket shell, this project confounds conventional assumptions about what makes a civic building of appropriate prestige and public face. The refurbishment project was a critical element of Moray Council's larger transformation programme, bringing council staff together onto a single city-centre campus and reducing its office estate from 21 properties to just five. Imaginative and rational design has resulted not only in a high-quality workplace, but also in significantly lower energy consumption than typical for the council's administrative functions.

CONTEXT AND OBJECTIVES

Moray Council employs 4,500 people, of which approximately 850 were located in 21 office and depot buildings dispersed throughout the Highlands cathedral city of Elgin. After reviewing the way services were delivered, it concluded that it needed to reduce and rationalise its current office space, institute flexible, mobile and home-working practices, and increase the use of electronic records and communications to reduce the burden of storing hard copies.

The council acquired a disused supermarket adjacent to its headquarters and in 2009, spotting its potential to accommodate these requirements, appointed a team led by Mace with Bennetts Associates as architect, to design the refurbishment. Amid a national economic depression and severe pressure on public finances, the council was very sensitive to the need to provide good value for money, and to be seen to be doing so. The brief was for uncomplicated open-plan offices, furnished with workstations and facilities that could accommodate new ways of working. The council wanted the building to serve as a base for most of their staff, requiring them to operate without dedicated workstations. The refurbished space also had to include a public reception area to give the people of Elgin a single point of access to council services.

Situated at the eastern end of Elgin High Street, next to Moray Council's existing council headquarters building, the site lies within a conservation area, about 200 m east of the pedestrianised section of the High Street, the main retail core of the city. The surrounding land uses include residential, commercial and retail.

Built in 1990, the original building served as a supermarket with a gross floor area of ca 3,000 m². Largely of concrete block construction with stone facings, it had a flat felt roof with slate details. The rest

01 Breakout area for informal meetings, with meeting 'pod' behind

02 Pre-retrofit interior

of the 2-acre site was given over to a 160-space tarmac-surfaced car park to the east of the building.

The council stipulated a BREEAM 'Excellent' rating as a contract condition and also asked for an EPC rating of 'B'.

DESIGN: GENERAL

While there are few external signs of change, the new entrance, sitting below a simple drum, appropriately signals the building's new purpose, with the previous tacky dome replaced by a minimal overhanging disc raised above clerestory windows. Upon arrival, the large dimensions of the entrance of the old supermarket and the adroit integration of disability ramps and steps make for an unusual generosity of space.

03 Light-filled workspace post-retrofit
04 Open-plan workspace with generous volume and daylighting
05 Pre-retrofit lobby
06 The new public entrance from the High Street replaces the former supermarket café
07 Pre-retrofit entrance area
08 The former entrance area has been opened up to provide a staff canteen with views out to the street

04

Inside, the feeling of spatial generosity is continued by the tall ceilings, with open-plan spaces easily accessible and clearly signposted using different colours: well populated, but uncluttered.

The existing supermarket was stripped back to its original structure, internally revealing an essentially unobstructed long-span steel-framed hangar. New saw-tooth windows were added to the stripped-back roof to provide ample levels of harmonious daylight to the inevitably deep plan and to provide natural ventilation, in combination with wind chimneys. The 5 m void between floor and roof is punctuated with effective acoustic baffles to mitigate noise levels and interference. The deep-plan layout straightforwardly separates public spaces from the private office space, the latter being clearly organised into colour-coded work areas, meeting rooms, breakout spaces and a canteen; all fit for flexible ways of working. The large informal breakout areas are interspersed throughout the 3,000 m² floorplate. There are a number of meeting rooms, and also two substantial training rooms, that can be combined to form a large-capacity Emergency Planning Room.

Completed in December 2011 and fully occupied since March 2012, the unostentatious scheme quietly rejuvenates a neglected, but important, corner of the city, unexpectedly turning an apologetically

09 Pre-retrofit exterior
10 The main entrance addresses Elgin High Street, offering public-facing access to council services

post-modern piece of architecture into a modest and dignified civic building.

DESIGN: ENERGY CONSERVATION

The design team prioritised simple measures; those that would produce the greatest reductions in carbon emissions for the least investment. Unsurprisingly, the design focused on passive engineering before active engineering – or in the words of the architect, plain 'good design' before 'good technology'. Beyond that, the council considered options for low- and zero-carbon renewable energy technologies. These were not adopted, but allowances have been built in to enable them to be added during a future upgrade.

By consolidating its building stock from 21 dispersed offices to just five centrally located ones, the council greatly reduced its per capita carbon emissions. The mere fact that this project reused an existing building dramatically reduced its overall environmental impact and contributed to its high BREEAM score. Coupled with this, new insulated linings to the external walls and new roof finishes improved the thermal envelope performance to 15% above Building Standards levels and ensured that the target air permeability of 5 $m^3/(h \times m^2)$ at 50 Pa was achieved.

Directional wind chimneys and windows in the reconfigured roof deftly combine a natural ventilation strategy with ample daylighting. Stripping out the existing finishes exposed the full 5 m-high volume of the building, making it possible to avoid the uncomfortable CO_2 build-up that typically arises in deep-plan offices. In this solution, stale air collects at the top of the space and exhausts through the wind chimneys. This exhaust is replaced with fresh air through new vents placed on the perimeter of the building.

During summer, the effect of wind on the directional chimneys creates a pressure differential that draws fresh air in at the perimeter and up through the space. On very warm, still days, this natural effect ceases to work, at which time fans kick in to assist. The same fans are used in reverse in winter months to blow heated air back into the building.

During winter, the chimneys are closed to prevent unnecessary heat loss. Instead, trickle ventilators are used, which rely on temperature differentials to function. Low-temperature hot water radiators positioned below the ventilators prevent the trickle being felt as a cold draught.

While inaccessible openings are controlled by the building management system, low-level vents are controlled manually by the people nearby. Internal meeting rooms are mechanically ventilated using local heat recovery.

A comparative analysis using low-velocity mechanical ventilation proved the viability of the natural ventilation scheme. Natural ventilation was estimated to generate emissions of 27 $kgCO_2/m^2/yr$, compared with 31 $kgCO_2/m^2/yr$ for mechanical ventilation (the council's other properties average 77 $kgCO_2/m^2/yr$).

Computer analysis helped to optimise the design of new rooflights. These have dramatically improved the amount of daylight, raising the average daylight factor to almost 5% in most working areas. The rooflights are north-facing with solid gables, avoiding the potential for glare or overheating.

USER BEHAVIOUR
AND BUILDING MANAGEMENT

Having been used to dedicated and highly serviced offices, council workers found that the move to an open-plan, low-energy building meant a lot of changes all at once. Needless to say, not only did the move require minute planning, it also called for significant change management. The refurbished building represents the new, transformed organisation. At the same time, it also affords the council the opportunity to give all staff a taste of the future; by providing decant space, it will enable the refurbishment of other campus buildings.

The council has an in-house facilities maintenance team, which controls the running of the building and whose records offer a good opportunity for objective data analysis. At the time of writing, the first year of raw data has become available (see below), and an interim project review is imminent. The council plans a full review once the campus transformation programme is complete in 2015. Anecdotally, evidence suggests the internal environment is well balanced and pleasant. Natural ventilation and high levels of natural light are ensuring both occupant comfort and lower energy demand.

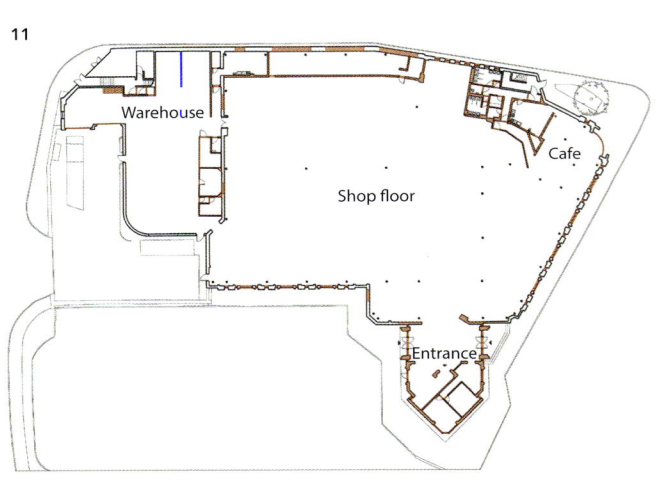

11

Warehouse

Cafe

Shop floor

Entrance

12

Moray Council HQ – Idealised environmental strategy

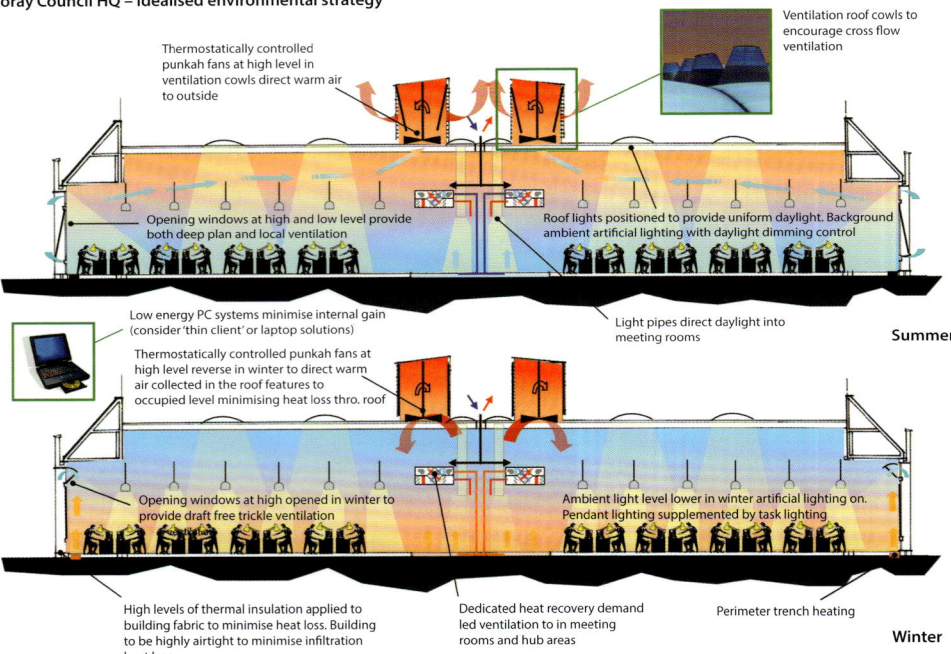

Thermostatically controlled punkah fans at high level in ventilation cowls direct warm air to outside

Ventilation roof cowls to encourage cross flow ventilation

Opening windows at high and low level provide both deep plan and local ventilation

Roof lights positioned to provide uniform daylight. Background ambient artificial lighting with daylight dimming control

Low energy PC systems minimise internal gain (consider 'thin client' or laptop solutions)

Light pipes direct daylight into meeting rooms

Summer

Thermostatically controlled punkah fans at high level reverse in winter to direct warm air collected in the roof features to occupied level minimising heat loss thro. roof

Opening windows at high opened in winter to provide draft free trickle ventilation

Ambient light level lower in winter artificial lighting on. Pendant lighting supplemented by task lighting

High levels of thermal insulation applied to building fabric to minimise heat loss. Building to be highly airtight to minimise infiltration heat loss

Dedicated heat recovery demand led ventilation to in meeting rooms and hub areas

Perimeter trench heating

Winter

11 Ground floor plan pre-retrofit
12 The environmental strategy
13 Long section pre-retrofit
14 Ground floor plan post-retrofit
15 Long section post-retrofit

13

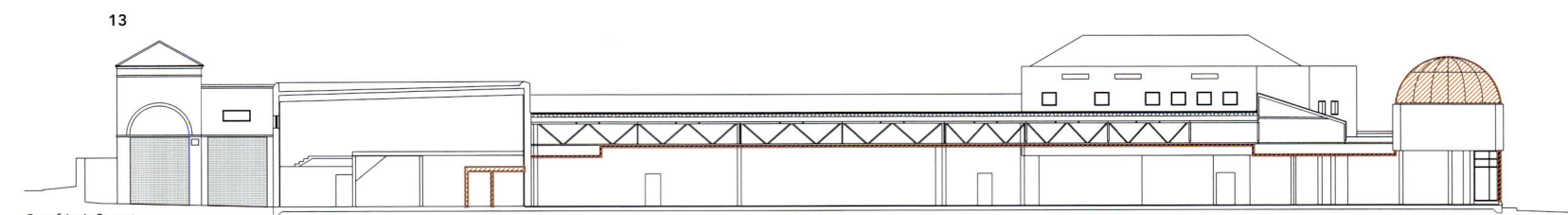

Greyfriar's Street

High Street

14

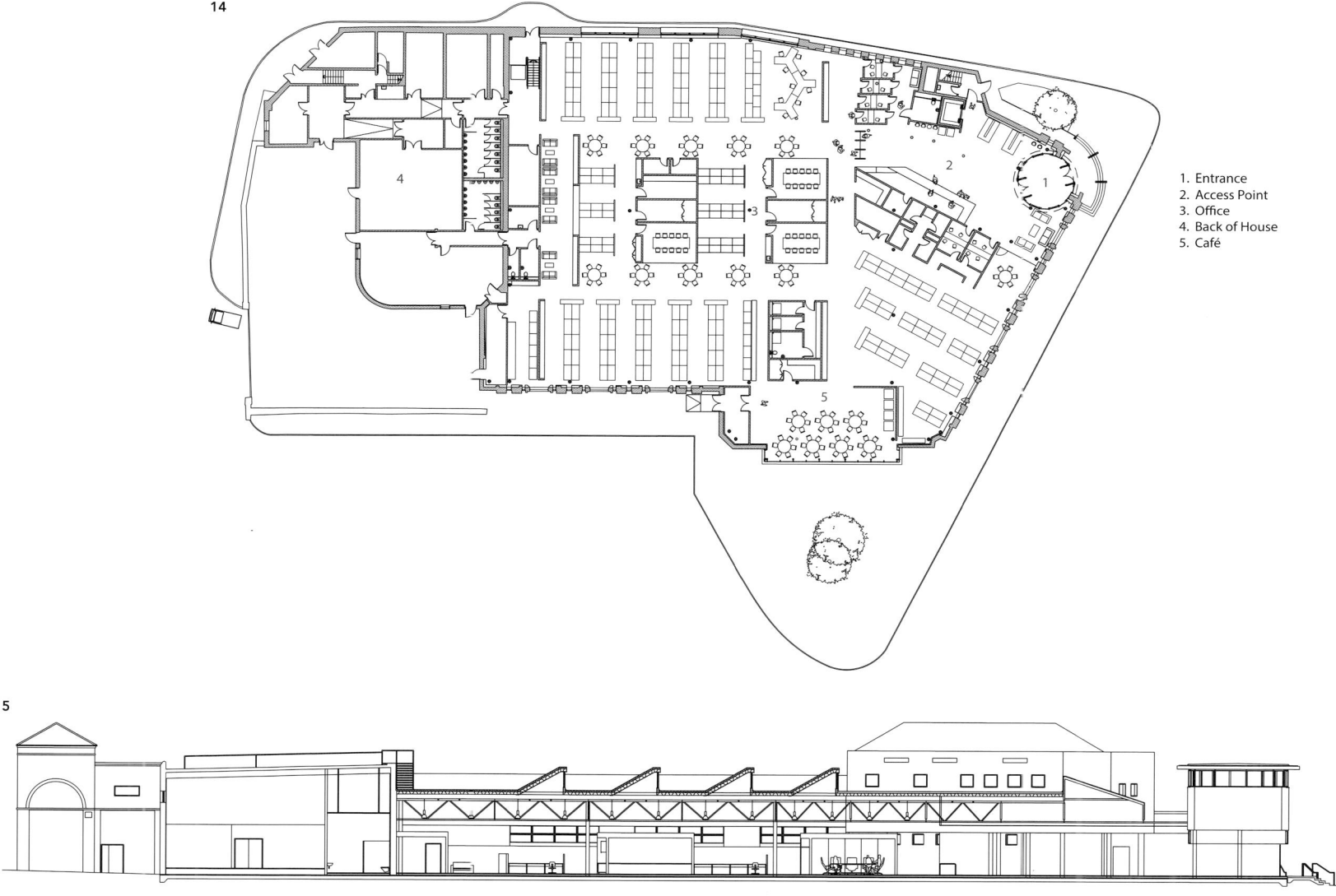

1. Entrance
2. Access Point
3. Office
4. Back of House
5. Café

15

Energy and carbon

MORAY COUNCIL HEADQUARTERS, ELGIN
Council headquarters

		Energy kWh/m²/yr	Carbon kg CO₂ / m²/yr	Building information		
				Area (m²)	Occupancy (m²/person)	Hours of operation
PRE-RETROFIT	Gas			2,943		
Monitored	Electricity regulated					
	Total regulated					
	Electricity unregulated					
	Total	n/a	n/a			
POST-RETROFIT DESIGN DATA				2.943	n/a	09:00-17:00
	Part L model	n/a				
	Predicted energy model					
	Gas	94.1	18.8			
	Electricity regulated	26.8	13.4			
	Total regulated	120.9	32.2			
	Electricity unregulated	45.5	22.8			
	Renewables	0	0			
	Total	166.4	55.0			
ACTUAL DATA	Post-occupancy data					
	Gas	132.9	26.6			
	Electricity regulated	41.4	20.7			
	Total regulated	174.3	47			
	Electricity unregulated	180	90.0	Data centre for whole council based here		
	Renewables	0	0			
	Total	354.3	145			
BENCHMARKS	Average emissions Moray Council Offices			Env. rating	EPC rating	DEC rating
	77 kg/m²/yr			BREEAM 'Excellent' (Design)	B	n/a
CONVERSION FACTORS	Gas	0.2				
kgCO₂/kWh	Electricity	0.5				

In this type of retrofit project it is almost meaningless to compare before and after energy figures; you cannot compare a supermarket's cooling-dominated energy use with that of an office in a cold climate. But for the record, the Energy Performance Certificate rating pre-retrofit was 'G' and post-retrofit is predicted to be 'B'.

What is more interesting and instructive is to compare the predicted and actual figures after one year of data. Overall, the energy performance of the refurbished building falls significantly short of the predictions, which is a common experience. However, a closer look at the numbers shows that the figures are heavily influenced by the fact that the council's entire data centre is located in the building, accounting for approximately 180 kWh/m²/yr of electricity use and carbon emissions of 98 kgCO₂/m²/yr, as much as all the other energy uses put together.

Looking at the energy use excluding plug loads, it was found that the first year's actual use was about 40% higher than the estimated use. The design team for the project have initially identified two possible reasons for this difference: the estimates relied on weather data from Glasgow, where winters are not as cold as in Elgin; and the winter covered by the actual dataset was particularly severe. There may well be other factors related to installation and commissioning; it is difficult to establish this as the design was completed by the contractor with the original environmental consultant having no delivery role.

Resources and waste

The reuse of a 20-year-old building was in itself a key component of the sustainability strategy. The analysis of a previous Mace and Bennetts Associates-designed project (see Elizabeth II Court for Hampshire County Council) showed that the reuse of an existing building frame (and here a large part of the fabric) reduces a project's embodied carbon by half. There was every reason to think that a similar saving could be made in this project.

The design team considered life-cycle costs to inform decision making, particularly important for a long-term investment by a risk-averse public body sensitive to the general climate of economic austerity. Evidence from the analysis revealed the optimum balance of environmental benefit and capital outlay. For example, comparative analysis showed that the natural ventilation scheme, which was subsequently adopted, offered a whole-life cost saving of £350,000 over a 25-year period, on top of its considerable saving in embodied energy, when compared with installing and maintaining a mechanical system over the same period.

Cost and time

Gross internal area:	2,943 m²
Approximate total construction cost:	£4.2 million
Cost per m²:	£1,330
Contract duration:	16 months
Completion:	December 2011

A project budget of £4.38m was established by the council, which funded the retrofit in full. This included all construction costs, furniture, surveys, applications for statutory consents and professional fees, but excludes IT hub/server equipment relocation, IT hardware and VAT.

Procurement

The project was designed and constructed in a single stage. It was procured using a design and build contract, with the design information prepared to RIBA Work Stage D before being carried forward by the contractor's team. The client team took steps to ensure that design-stage sustainability strategies and obligations were communicated explicitly throughout the procurement process.

The council's pre-qualification documents made it clear that contractors must understand and have experience of delivering low-energy buildings. Equally, the design specifications in the tender information emphasised the contractor's obligations in respect of building performance. For example, the building had to achieve a 'B' rated Energy Performance Certificate, specific levels of daylighting and airtightness, minimum U-values and target water consumption, and a BREEAM 'Excellent' rating.

Further quality fail-safes included arranging mid-tender meetings with all the tendering contractors to explain the design team's design intentions and sustainability objectives. Question-and-answer sessions allowed the contracting team to probe the design's underlying emissions projection assumptions.

The form of contract used was SBCC Design and Build (without novation). Bennetts Associates had no inspection duties but were retained in a consulting role for times when the client or the employer's agent wanted the architect's opinion on the contractor's proposals or variations to the contract.

16 'Pod' meeting space

CREDITS

CLIENT
Moray Council

PROJECT MANAGER
Moray Council

DESIGN AND BUILD CONTRACTOR
Stewart Milne Construction

QUANTITY SURVEYOR
Mace pre-contract and
Moray Council post-contract

**TECHNICAL ADVISER/
EMPLOYER'S AGENT**
Mace

ARCHITECT
Bennetts Associates

IMPLEMENTATION ARCHITECT
Acanthus Architects df

**STRUCTURAL AND
SERVICES ENGINEER**
Buro Happold

CDM CONTRACTOR
Mace

PHOTOGRAPHY
Keith Hunter and
Douglas Gibb

Percy Gee Building, Students' Union, Leicester

UNIVERSITY OF LEICESTER

This is more than a strictly 'retrofit' project: it involved the partial redrawing of the building envelope to provide the large-scale spaces needed for modern university purposes which the 50-year-old building simply did not have; these new spaces make the existing accommodation work far better. At about half the cost of a new-build and generating energy savings into the future, the project demonstrates what a client with clear goals for design quality and sustainability can achieve when working with designers able to apply a little lateral thinking alongside expertise.

01 Nightclub at the students' union
02 Nightclub pre-retrofit

CONTEXT AND OBJECTIVES

The Percy Gee Building is one of the larger buildings on the University of Leicester campus and houses the students' union. It is a strategically important building for the client, playing a central role in attracting and retaining students and staff.

Originally designed by T. Shirley Worthington as a social facility for a rather smaller institution, it was opened by the Queen in 1958, the year after the college became a university. Photographs show an architecture that combined Scandinavian modernism with neoclassical style in a robust brick exterior housing elegant, uncluttered spaces.

U-shaped in plan, the building sits to one side of the main entrance of the main campus. It is cut into the sloping site, with entrances at different levels on the north, west and south facades. As was then usual, there was little consideration of the needs of people with disabilities, with 26 different floor levels over five storeys and no lifts.

By 2008, after five decades of ad hoc adaptations, the building had become shabby and disorientating, was poorly regarded and a target for demolition. A review of the campus, which led up to a masterplan proposing a total investment of £1bn, at first considered a new-build replacement located in the centre of the campus, on a site to the east of the existing building. However, further analysis revealed that of all the uses which the vacated building might be considered for, a students' union was the most suitable, assuming the building's major shortcomings could be addressed.

The project was steered by a project implementation team, which sought to ensure that resources dedicated to the project were used in pursuit of the wider academic and social objectives of the university. The primary objective was 'the improvement, conversion

and extension to secure the long-term future success of the University and ensure that its students are offered a quality of facilities and experience appropriate to the University's status as one of the top twenty higher education institutions in the UK'. The brief asked for the building to be easier to use and more attractive, while aiming for a BREEAM 'Excellent' rating with an 18% contribution to energy from renewable sources – all in line with the university's policy.

03 Perspective drawing 1958
04 Plant and bridge link on the edge of The Square pre-retrofit
05 New atrium slot on the edge of The Square linking all levels
06 The building from University Road pre-retrofit
07 The students' union from University Road post-retrofit
08 The Square pre-retrofit
09 The Square from the balcony post-retrofit

DESIGN: GENERAL

The key move in the design strategy was to fill in the unused courtyard formed by the U-shaped plan, replacing rather grim flat roofs and mechanical plant with a tall atrium – 'The Square' – surrounded by galleries and bridges that create links with and between the existing spaces. These galleries and bridges unobtrusively accommodate the changes in level while lending theatricality and animation to the whole ensemble, from the cafeteria and social space on the ground floor to the various office, work and other social spaces that overlook it. They also achieve the integration of access for people with and without disabilities. Alongside the main atrium a second narrow and taller four-storey atrium running north–south reaches down to basement level, bringing light into a previously dark circulation spine and creating yet more dramatic transparency.

The glazed atrium roof is supported on wide-span curved glulam timber beams in a wishbone pattern, which, in turn, sit on steel trees, which suit the structural arrangement of the floors below. The approach to this new space from outside is by a grand external staircase leading up to a west-facing paved terrace that overlooks the mature trees in the Victorian cemetery opposite and creates a new front that addresses an important direction of approach. The size and visual connection to the surrounding area create much improved urban relationships for an important university building.

The original basement floor under The Square was dug out and lowered to expand and improve the existing nightclub, creating a well-appointed and highly subscribed venue for the university and, indeed, the whole of Leicester. Reached from the lowest floor of the slot atrium, it now has three performance rooms and its capacity has been increased to 1,750. The stage has been relocated to the west end, closer to new dressing rooms that were inserted over a remodelled service yard, tucked under the new terrace. These changes provide increased capacity for touring bands and their coaches and articulated lorries.

10 Concourse in 1958
11 Concourse pre-retrofit
12 Concourse post-retrofit

A major consideration for the client was noise breakout from the nightclub, both for the health and well-being of users and for good neighbourliness. The space is therefore highly insulated and attenuated. The ventilation equipment is housed under the stage, designed to take in fresh and exhaust stale air without sound transmission to the outside.

A number of fine original features and spaces of the existing building have been saved and revealed. For example, a lift that obscured the cantilevered helical main staircase has been removed following the addition of three new lifts. Timber classical columns and plaster cornices, discovered when the many generations of bar fit-outs were stripped out, have been retained and restored. The transformed 8,500 m^2 building also accommodates a music venue, shops, offices, restaurant, bar, coffee bar, resource centre, welfare and clinic facilities, a gym and a meeting rooms suite. All 276 rooms or spaces in the building have been improved.

DESIGN: ENERGY CONSERVATION

The design incorporates a pragmatic energy strategy. Natural ventilation plays a big part, with large ducts prominently integrated into the spatial design. These ducts allow the nightclub, with its very demanding conditions, to be naturally ventilated, with fan back-up but no cooling. During a nightclub crush, CO_2 sensors trigger a back-up mechanical ventilation system to increase the air supply. The thermal mass, both of the existing building and of the heavy concrete structure

and floors of the new construction, combined with the stack effect in the atria, is used to keep the main highly populated areas cool. Ventilation louvres in the west elevation and at roof level are automatically operated by the building management system, which senses temperature and CO_2 levels and adjusts the amount of fresh air flowing through the space. In summer, much of the west face can be opened up onto the terrace, allowing the breeze to circulate to help keep the building cool, while fixed external louvres over the large glazed areas help control solar gain. The older parts of the building have no significant cooling demand as yet.

As for heating loads, the existing fabric has not been insulated, but heat loss from the walls facing the old courtyard is greatly reduced because of the atrium, which is itself built to relatively low U-values despite the glazing. This glazing in turn minimises artificial lighting loads. The main active measure is a biomass boiler that uses locally sourced woodchips, which accounts for the major part of the reduction in carbon emissions. There is also a back-up gas boiler to ensure continuity in case woodchip supply is ever disrupted.

Other environmental measures include:

- the retention of the existing structure and fabric (and thus its embodied energy)
- specification of sustainable new materials
- energy-efficient lighting throughout, controlled by movement sensors
- Forest Stewardship Council certified new timber

- rainwater harvesting from the roofs, with water storage in the basement and under the service yard for use in low-flush toilets.

USER BEHAVIOUR
AND BUILDING MANAGEMENT

Two years after completion, the number of visitors had tripled and the building had recorded 12,000 visits in a single week. With these numbers, user control of the environment is impractical and energy conservation relies on automation and central management.

A post-occupation evaluation carried out in spring 2012 concluded that after a complex construction project, which was carried out with staff and students remaining in occupation throughout, the building was performing well and the client was very satisfied. The evaluation reported many positives from the project: the space worked well as somewhere for students to work or socialise; it helped the university to promote itself and its outward physical presence; it greatly improved accessibility; its green credentials are good; and it has cut down the university's backlog of maintenance. Almost all problems reported by users were already being dealt with as a matter of post-completion liaison between the design team and the university.

13

3

4

4

2

5

1

1 Terrace over new nightclub
2 The 'Square' central heart created in un-used courtyard
3 New slot atrium connecting all levels with walkways
4 Ventilation chimneys for nightclub
5 East wing extended

13 Computer model aerial view of
   the students' union building
14 Level 3 plan pre-retrofit
15 Long section pre-retrofit
16 Level 3 plan post-retrofit
17 Long section post-retrofit

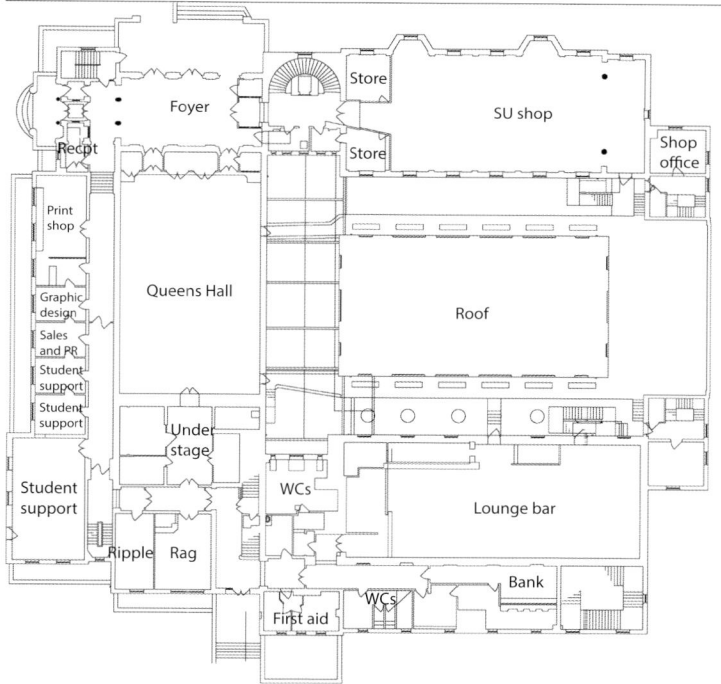

Recpt
Foyer
Store
SU shop
Store
Shop office
Print shop
Queens Hall
Roof
Graphic design
Sales and PR
Student support
Student support
Under stage
Student support
WCs
Lounge bar
Ripple
Rag
First aid
WCs
Bank

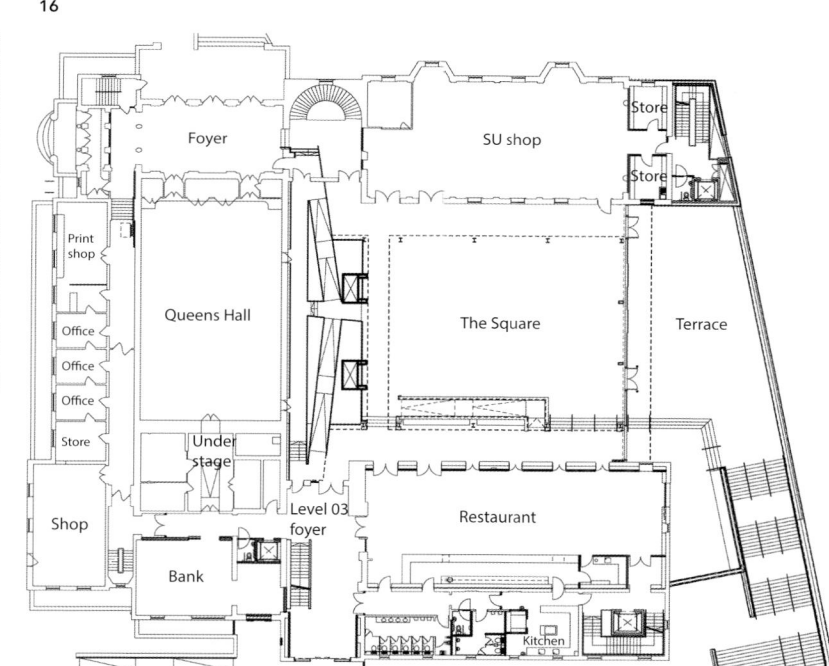

Foyer
Store
SU shop
Store
Print shop
Queens Hall
The Square
Terrace
Office
Office
Office
Store
Under stage
Level 03 foyer
Restaurant
Shop
Bank
Kitchen

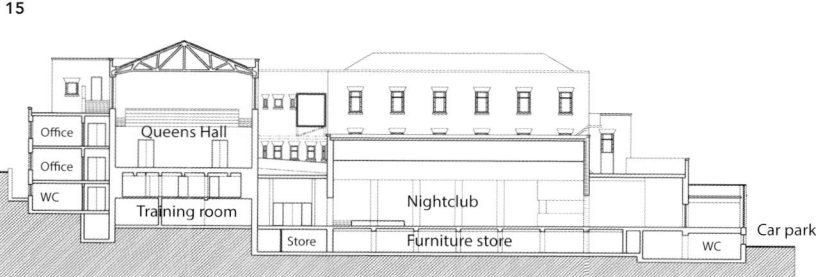

Office
Office
WC
Queens Hall
Training room
Store
Nightclub
Furniture store
WC
Car park

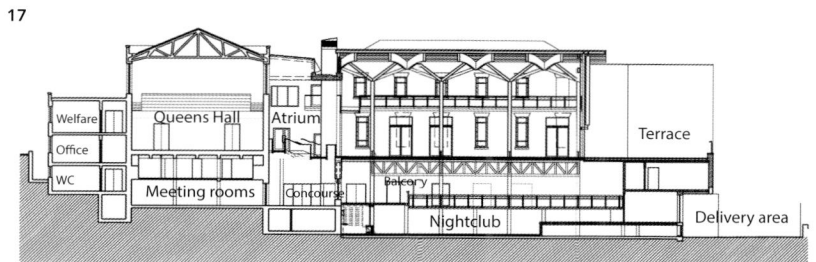

Welfare
Office
WC
Queens Hall
Atrium
Terrace
Meeting rooms
Concourse
Balcony
Nightclub
Delivery area

DATA

Energy and carbon

PERCY GEE BUILDING, UNIVERSITY STUDENTS' UNION
University students' union

		Energy kWh/m²/yr	Carbon kg CO$_2$ / m²/yr	Building information		
				Area (m²)	Occupancy (m²/person)	Hours of operation
PRE-RETROFIT	Gas	180.19	36.04	8,101	2.15	08:00–02:00
Averaged over 2004–2008	Electricity regulated	60.22	30.11			
	Total regulated	240.41	66.15			
	Electricity unregulated	inc.	inc.			
	Total	240.41	66.15			
POST-RETROFIT DESIGN DATA				9,804 GIA 8, 500 Net	1.81	08:00–02:00
	Part L model (2006)		51.15			
	Predicted energy model					
	Gas	0	0.00			
	Electricity regulated	102.78	51.39			
	Total regulated	102.78	51.39			
	Electricity unregulated					
	Renewables	98.61	0.00			
	Total	201.39	51.39			
ACTUAL DATA	Post-occupancy data					
	Gas	99.59	19.9			
	Electricity regulated	99.63	49.8			
	Total regulated	199.22	69.7			
	Electricity unregulated	inc.	inc.			
	Renewables	11.04	0.0			
	Total	210.26	69.7			
BENCHMARKS	CIBSE Guide F	320		Env. rating	EPC rating	
				BREEAM (2008)	B (28)	New parts
				'Excellent' (Design-target not certified)	B (46)	Old parts
CONVERSION FACTORS	Gas	0.2				
kgCO$_2$/kWh	Electricity	0.5				
	Wood pellets	0				

Energy Performance Certificates from before and after the refurbishment indicated that carbon emissions across the building would be lowered by 20–40%. However, initial in-use figures show that this performance is not yet being achieved. This is most likely to be due to higher electrical loads from the more intensive use of the building. The natural ventilation system for the night club, the reduced heat loss through walls that were external and are now internal, and the biomass boiler must be reducing regulated loads. The predicted energy figures relate to a design which has been significantly altered and are therefore not useful for analysis. Energy and heat monitoring systems were installed as part of the retrofit, and the building management system can be monitored and controlled remotely. Interrogation of the energy figures is continuing.

Resources and waste

The university's policy is to achieve 75% recycling (by weight) over the next ten years, through 5% biannual increases, and to aim for a 1% decrease in waste figures per capita. For this project, a survey was carried out prior to construction to review how much of the building fabric and fittings could be reused or recycled. During construction the main contractor used sustainable waste management disposal and provided records of the types and proportions of materials disposed of.

The design of the building enables the students' union and university to optimise the amount that it recycles, and to store materials for composting. A dedicated segregated space for storing compostable food waste is provided within the service area of the building as part of the university's campus-wide programme of collection and removal off site for composting. A compactor is located in the service yard to reduce the bulk of other waste before it is transported off site.

Cost and time

Gross internal area:	8,500 m²
Construction cost:	£12.65m
Cost per m²:	£1,488
Project programme:	
Design team appointed:	April 2008
Building contractor appointed:	December 2008
Planning permission received:	December 2008
Building began:	May 2009
Completion:	November 2010
Contract duration:	18 months

Procurement

The design team was appointed in April 2008. A two-stage tender process allowed the contractor to investigate and understand the existing building thoroughly, particularly the mechanical and electrical services, before submitting second-stage tender proposals. The form of contract used was JCT Standard Building Contract Without Quantities 2005 edition (Revision 1 2007), with contractor's designed portions for precast concrete stairs, steelwork connections and glulam beam connections. The contract works started on site in May 2009.

The 64-week main contract works were arranged in six phases to allow the building to remain occupied throughout. Fire escape routes had to be adjusted between phases, requiring close liaison between contractor, the students' union, building control and estates officers.

The scope of the works was increased and varied by the university during the course of the contract as additional funds became available through fundraising, and the university carried out its own direct works in parallel with the main works.

The main body of the works, worth £12.65m, was completed in September 2010. With the university's direct works, the overall project cost (including VAT) was £17m.

CREDITS

CLIENT
The University of Leicester and University of Leicester Students' Union

PROJECT MANAGER
Currie and Brown

MAIN CONTRACTOR
Morgan Sindall

COST CONSULTANT
MDA

BREEAM FACILITATOR
Code Green

BUILDING INSPECTOR
Leicester City Council

ARCHITECT
Shepheard Epstein Hunter

STRUCTURAL ENGINEER
Scott White & Hookins

ENVIRONMENTAL ENGINEER
WYG

FIRE ENGINEER
FDS Ben Whitaker

CDM COORDINATOR
Faithful and Gould

PHOTOGRAPHY
Martine Hamilton Knight

Mildmay Community Centre, London

LONDON BOROUGH OF ISLINGTON

This project is the first UK example of the Passive House standard being applied to the refurbishment of a non-domestic building. Passive House design techniques aim to produce comfortable and healthy buildings that use much less energy than a normal building. This building has been made more comfortable while also delivering measured annual energy savings of 80%. It has also become more spacious, usable, accessible and delightful.

02

01 Exterior view across the garden
02 Garden view pre-retrofit

CONTEXT AND OBJECTIVES

The community centre, located in one of London's most deprived areas, was in urgent need of total renovation. The original building, a late 19th-century electricity generating station, was rescued from dereliction in 1973 by the Mildmay Community Partnership. But it had poor accessibility, and being uninsulated it was cold; only 60% of the floor area was usable due to the cold draughts that swept through the basement. By 2006, Islington Council had adopted a firm sustainability agenda, strongly supported by elements within the leadership of the community centre, such as the chair of the Sustainability Committee, Jenny Littlewood. When the possibility arose of refurbishing the building it was decided to vigorously tackle energy efficiency, both on the grounds of reducing waste and 'showing the way'.

Later that year the client asked three architects to present their proposed approaches to a committee of building users. bere:architects, whose founder Justin Bere was a local resident, was selected to refurbish the rundown building on the basis of their approach, which included changing the orientation of the building, to enjoy a garden to its south; creating 30% additional useable space in the basement, by adding windows providing daylight via a south-facing excavation; improving the comfort and accessibility of the building; and offering the potential to reduce energy consumption by an estimated 80%.

DESIGN: GENERAL

The additional usable space generated through converting the basement for use by both the local community and small businesses was crucial for increasing revenue potential. A single-storey extension was demolished and rebuilt, filling in a small courtyard to form a new entrance, reception and dining area.

The new step-free entrance, level floor surfaces and electrically controlled entrance door made the building accessible to all. Community and stakeholder consultation informed the design, and events organised by the Mildmay Community Partnership and the architects kept residents informed about the proposed redevelopment and gave them a platform to express their needs and views.

There is something just right about the retrofit of the Mildmay Centre. It's not just that the place feels serene and welcoming and the spaces are flooded with daylight; it's the quality of the 'haptic moment', when the body engages with the building. The triple-glazed front door feels solid, and it closes firmly. The openable triple-glazed windows are draught-free, even in winter, and their internal surfaces are completely free from condensation all year; they are almost as warm as the internal wall surfaces, which maintain a steady 20–21°C. This is part and parcel of meeting the Passive House standard. It is as much a comfort standard for temperature and air quality as a low-energy retrofit solution.

The understated architectural language of glazed partitions, plywood and exposed services helps legibility, as does careful design of simple control

panels, with their open/closed, off/on options. The Mildmay Centre is now an urban village hall, used by local clubs and societies and for weddings and celebrations. It promises to act as a catalyst for the regeneration of the surrounding community, while serving its practical needs.

DESIGN: ENERGY CONSERVATION

It was decided to adopt the Passive House approach in 2006, three years before the first Passive House was built in the UK. There were already numerous Passive House buildings in Europe following the successful EU-funded pilot building trials approximately ten years earlier. This project is the first example of its use for a non-domestic refurbishment project in the UK.

03 Exterior view of new dining area
04 The building pre-retrofit
05 Interior of the new dining area
06 External insulation being installed
07 The garden post-retrofit

Passive House is a demanding construction standard for very low energy buildings. A key objective is to achieve a very high level of indoor comfort and health, while using very little heating or, in warm climates, cooling: it sets a maximum 'specific heat demand' of 15 kWh/m²/yr. There is also a limit of 120 kWh/m²/yr 'primary energy demand', which is a measure that takes into account losses due to fuel conversion in a power station. This energy must cover hot water, heating and cooling and auxiliary and household electricity.

The standard can be defined as follows:

> A Passive House building has an extremely small heating energy demand and therefore needs no active heating. Such buildings can be kept warm 'passively' by using the existing internal heat sources and the solar energy entering through the windows as well as by the minimal heating of incoming fresh air, using heat extracted from outgoing stale air.

In practice this means eliminating the need for a conventional heating system. A Passive House building requires a draught-free building envelope and uses a heat recovery ventilation system, which extracts most of the heat energy from the stale outgoing air via a heat exchanger (without mixing the air) and uses it to warm the fresh incoming air.

Measures to upgrade the fabric comprised wrapping the building in external insulation, replacing the windows with triple-glazed units, installing large south-facing windows with retractable solar blinds and virtually eliminating air leakages. Active measures comprised an 18 kWp photovoltaic array, a solar thermal panel and an 8 kWp ground source heat pump (smaller capacity than a domestic boiler) capable of supplying a small amount of low-temperature heat to the internal radiator circuit. No radiators were installed in the basement – subsequently, the minimum temperature recorded in the basement offices, on a Monday morning in 2013, during the worst UK winter for 40 years, was 19.75 °C.

Passive House methods require careful construction and commissioning of the building. The envelope was pressure-tested early in the construction process, when the membranes were accessible and could be remedied to achieve the demanding standard of less than 0.6 air changes per hour at 50 Pa pressure difference between the inside and outside air. This was done after installing the windows and doors but before installing insulation and external cladding. This kind of attention to detail in planning and construction is essential if the demanding design objectives are to be met.

Other energy conserving measures include: 100% low-energy fluorescent lighting, controlled manually, but also provided with passive infrared presence sensors in case lights are left on; microbore hot-water pipes, to minimise standing pipe losses; high-quality insulation to all hot-water pipework; and external solar shading, which combines with secure night-time natural ventilation to control unwanted summer heat gains. Turn-and-tilt windows, openable rooflights with rain sensors and an insulated ventilator with external security grille all help provide natural ventilation in summer.

USER BEHAVIOUR AND BUILDING MANAGEMENT

As with all buildings, a Passive House one will perform best if its users understand how to get the best out of the building, but there is evidence that Passive House is an unusually robust way of delivering designed energy targets. The constant flow of fresh air makes opening a window in winter feel unnecessary, and the lack of cold draughts stops the thermostat being turned up in compensation.

The two spaces of very variable occupancy (the main hall and the dining area) have CO_2 sensors that automatically increase the air supply if the CO_2 level reaches 1000 ppm, thereby maintaining substantially healthier levels than required by national and European guidelines.

The building's low energy demands have meant that the controls are as simple as in a domestic house and so there is no need for a complex and expensive building management system. All controls are intuitive and well labelled. A single room thermostat in the main hall, which is independent from the other controls, calls for heat from the ground source heat pump if necessary. A time switch controls the heat recovery ventilation unit, turning it on in the morning and off again at night, with a boost button for use outside normal hours of occupancy. The CO_2 sensors in the main hall and dining area ensure excellent indoor air quality even when the hall is full of people.

The building has a diverse and ever-changing clientele and so managing occupant behaviour – e.g. stopping doors being left propped open – requires constant attention. The Soft Landings aftercare programme for building handover and occupant operation has been implemented. It encourages and helps the correct commissioning of a building's systems. It also ensures that design strategies are explained to users and that performance results are fed back to users to encourage them to take responsibility for energy use.

The continuing involvement of the design team and contractor deters wasteful practices and promotes a verifiably sustainable operation of the building.

08 The main hall post-retrofit

09

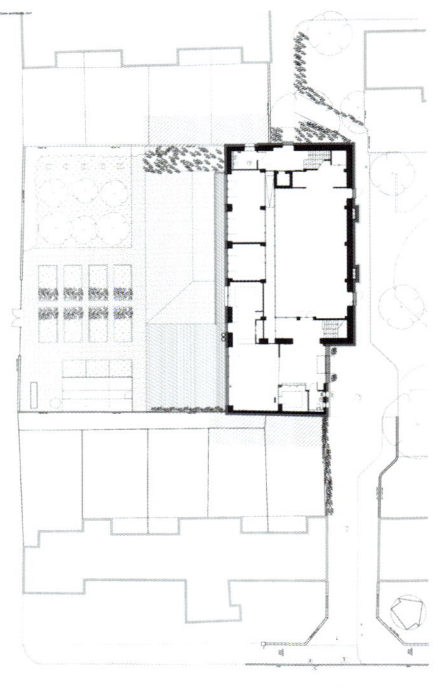

10

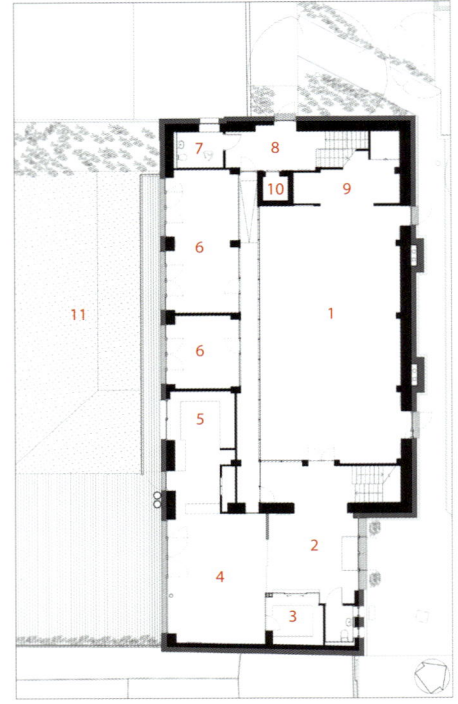

1 Hall (125 m²)
2 Foyer
3 Reception
4 Dining area
5 Kitchen
6 Office space

7   WC and baby change
8   Lobby
9   Store
10 Lift
11 Turfed grass area

11

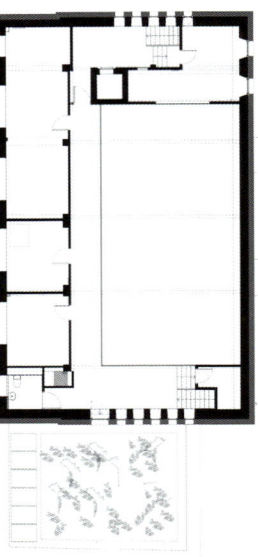

09 Site plan

10 Ground floor plan

11 First floor plan

12 Cross-section

1 Fixed-triple glazed rooflights

2 Solar panels

3 Photovoltaic panels

4 Existing brickwork wall externally insulated and rendered using Permarock insulated render system – total additional build-up 300mm

5 Zinc standing seam roof

6 Zinc standing seam finish to copings and faces of parapets

7 New warm roof insulation build-up with ecological roof garden

8 Triple-glazed Passivhaus certified timber windows

9 New balcony to ground floor offices

10 Junkers sprung timber floor to main hall

11 Excavated and installed 200mm insulation to external face of basement wall, installed french drain and back-fill with foam glass granules
100 mm Styrofoam Floormate A (inner layer of insulation)
100 mm Styrofoam Perimate (outer layer of insulation)

12 Exact extent of existing foundation unknown

13 New floating timber floor to basement

14 Photovoltaic panels on steel framework

15 New fixed window to first floor office spaces

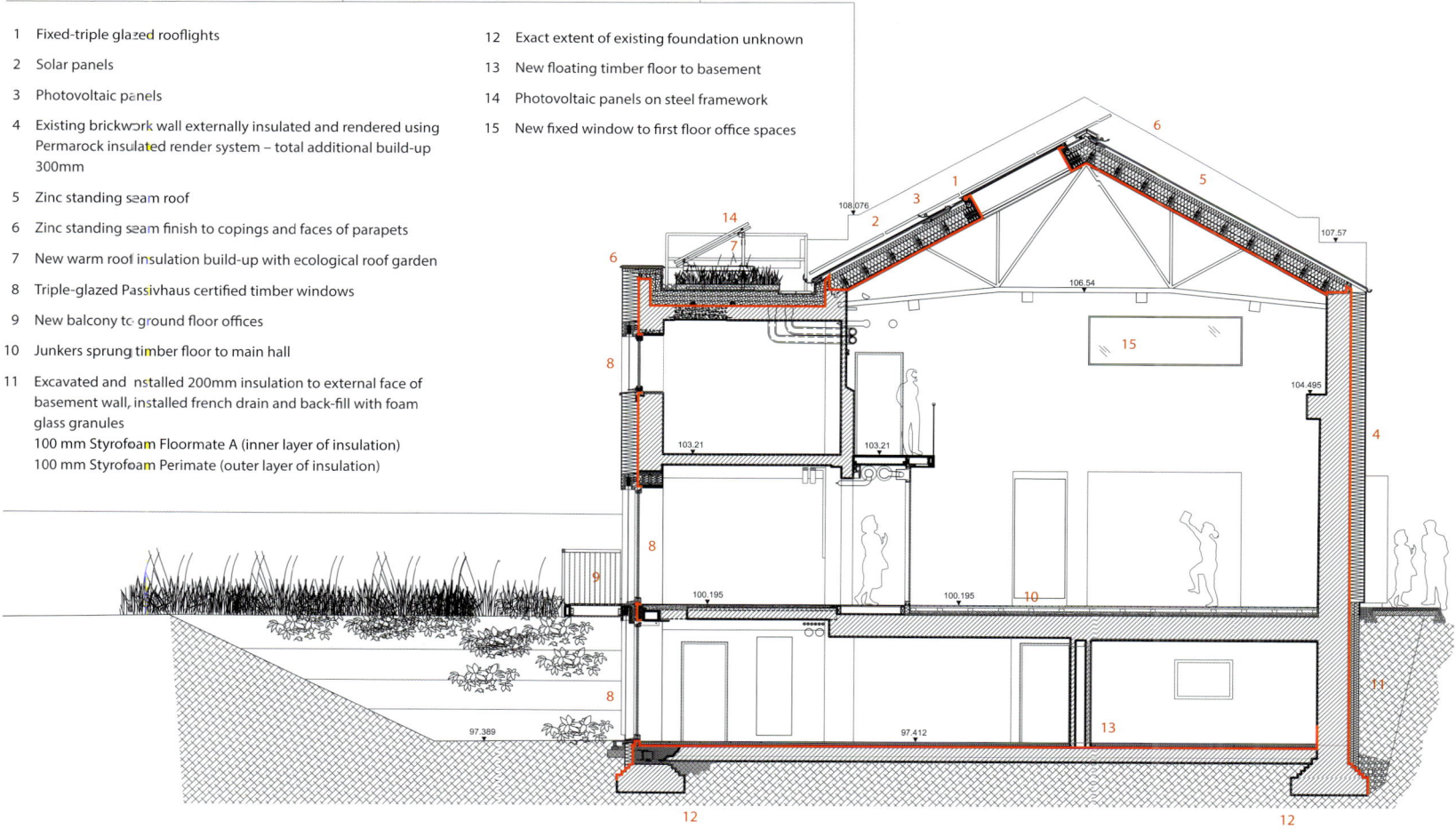

DATA

Energy and carbon

MILDMAY COMMUNITY CENTRE
Community centre with hall, café and garden

		Energy kWh/m²/yr	Carbon kg CO₂/m²/yr	Area (m²)	Occupancy (m²/person)	Hours of operation
		Energy $kWh/m^2/yr$	Carbon $kg\,CO_2/m^2/yr$	Area (m^2)	**Building information** Occupancy $(m^2/person)$	Hours of operation
PRE-RETROFIT	Gas				n/a	08:00-22:00
Monitored	Electricity regulated					
	Total regulated					
	Electricity unregulated					
	Total (est.)	580				
POST-RETROFIT DESIGN DATA				800 (GIA)		
				665 net		
	Part L model SBEM		13.7	This figure based on 800 m²		
	Predicted energy model					
	Gas	0	0	These figures based on 665 m²		
	Electricity heating (GSHP)	5.6	2.8			
	Electricity lighting	15	7.5			
	Electricity/total	24.3	12.2	Includes GSHP, DHW and auxiliary electricity (pumps, etc.), lighting		
	Renewables: PV	-21.7	-10.9			
	Renewables: ST	-2.0	-1.0			
	Electricity unregulated	20.2	10.1	Calculated from appliance ratings		
	Total	20.8	10.4			
ACTUAL DATA	Post-occupancy data					
(2012/13)	Gas	0.0	0.0			
	Electricity/total	16.5	8.3	Total electricity minus calculated unregulated loads		
	Renewables PV generated	20.1	10.1	Metered 12 months' data		
	Renewables PV exported	4.1	2.0	Sub-metered and calculated data		
	Renewables PV used	16.1	8.0	PV generated minus PV exported		
	Total import from grid	50.5	25.2	Sub-metered data		
	Electricity unregulated	50.0	25.0	Calculated through equipment hours of use and ratings		
	Total	66.5		Sub-metered and calculated data, PV used + grid import		
BENCHMARKS				**Env. rating**	**DEC rating**	
				Passive House	n/a	
CONVERSION FACTORS	Gas	n/a				
kgCO₂/kWh	Electricity	0.5				

Notes

1 The Passive House methodology employs the net figure for energy calculations – in this case 665 m²

2 Passive House uses a conversion of factor of 0.68 kgCO₂/kWh.

3 Using the GIA figure and a conversion factor of 0.5, energy and carbon in-use calculate as follows:
Total import from grid:
Energy 41.9 kWh/m²/yr
Carbon 21.0 kg CO₂/m²/yr

4 Using Passive House methodology – net usable area figure and a conversion factor of 0.68 – energy and carbon in-use calculate as follows:
Energy 50.5 kWh/m²/yr
Carbon 34.3 kg CO₂/m²/yr

5. GIA, gross internal area; GSHP, ground source heat pump; PV, photovoltaics; ST, solar thermal.

The building meets and exceeds all current applicable UK regulatory energy standards. It also exceeds Building Regulations Part L2A, which applies to new non-domestic buildings. SBEM (Simplified Building Energy Model) calculations gave the following results: against the target CO_2 emission rate (TER) of 18.8 $kgCO_2/m^2/yr$, the predicted building CO_2 emission rate (BER) was 13.7 $kgCO_2/m^2/yr$.

Prior to refurbishment, the total energy consumption of the building was approximately 580 $kWh/m^2/yr$, with an annual energy cost at the time of £10,700. The proposed one was modelled using PHPP (Passive House Planning Package), which predicted an 80% overall reduction in energy use. Of this only 12 $kWh/m^2/yr$ would be used for space heating, with overall costs of less than £1,000 per annum at pre-retrofit occupancy and energy costs.

Independently verified monitoring is showing that the building is meeting its expected performance. By early 2013, a full year of data had been collected, a period that included a particularly severe winter. A comparison between the energy consumption before the retrofit works (gas and electricity bills from January to December 2009) and the energy consumption after refurbishment to an all-electric building (sub-metering data gathered throughout 2012) indicates a reduction in total energy consumption of over 80%, which is better than predicted. The high quality of the design and construction and the adoption of Soft Landings are the key factors in this achievement.

13

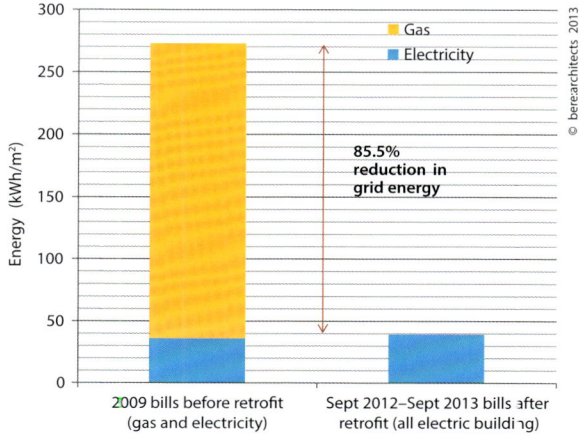

14

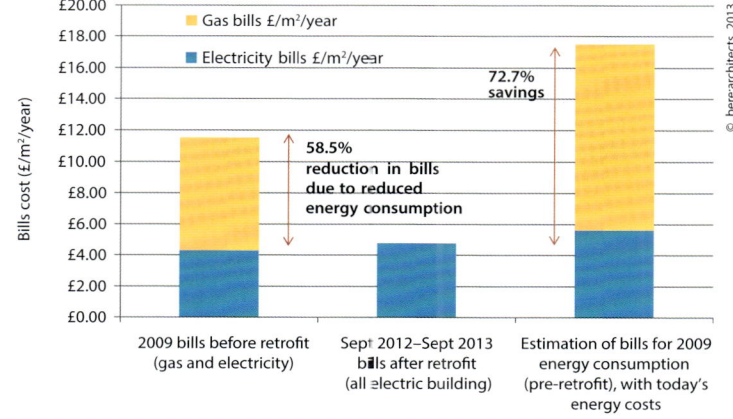

13 Grid energy consumption before and after retrofit
    – data from energy bills
14 Energy costs before and after retrofit

Although the building is meeting its expected performance, the unregulated loads – the energy used by plugged in appliances over which the designers have little control – are more than double what was expected. The energy table shows that the demand on the grid almost exactly matches the unregulated loads – i.e. the building is otherwise self-sufficient in energy – equating to zero carbon under the current official definition.

15 The reception area post-retrofit

This energy reduction was achieved while providing very comfortable internal conditions, as shown by the internal temperatures and relative humidity monitoring data. Collection of two years' performance data via the Technology Strategy Board Building Performance Evaluation Programme is providing evidence that this approach is an effective way to achieve big reductions in carbon emissions and energy use.

The definition of floor area in the PHPP excludes stairs and circulation spaces, therefore to calculate kWh/m^2 and $kgCO_2/m^2$ the totals are divided by $665 m^2$ rather than $800 m^2$. Additionally, the package uses a conversion coefficient of $0.68 kgCO_2/kWh$, rather than the best fit current official figure of 0.52. If the full floor area and the official conversion coefficient are used, the total annual actual net energy use is $42 kWh/m^2$ and the carbon emissions figure is $21 kgCO_2/m^2$. Summary of renewable energy systems:

- 77 x grid-connected NU235E1 Sharp photovoltaic array; $116 m^2$ total delivering 18 kWp (14,400 kWh)
- 1 x Viessmann Vitosol 200 solar panel system; $3 m^2$ delivering 3 kW (1341 kWh)
- 8.4 kWp Viessmann Vitocal 300-G ground source heat pump with 140 m double-pipe trench
- two rainwater harvesting tanks; collecting 11,000 litres of rainwater for WC flushing and garden irrigation.

Resources and waste

The project promoted a fabric-first approach to energy reduction. Saving the 450 mm-thick, solid brick walls minimised the embodied energy of the project as well massively reducing demolition waste.

Cost and time

Construction cost:	£1.6 million
Cost per m^2:	ca. £2,000
Start on site:	May 2010
Completion:	August 2011

Grant funding was available from Islington Council for renewables (photovoltaics and the ground source heat pump) and also for the heat recovery ventilation unit. The Carbon Trust, The Big Lottery Fund and the Community Builders Fund, as well as anonymous donors, also helped fund the project.

Procurement

The retrofit works were carried out under a 'traditional' contract using the JCT Standard Building Contract with Quantities (2005) form.

CREDITS

CLIENT
Mildmay Community Centre

CONTRACTOR
Buxtons

COST CONSULTANT
Richard Whidborne

GREEN ROOF CONSULTANT
Dusty Gedge

POST-OCCUPANCY INVESTIGATOR
Roderic Bunn

ARCHITECT
bere:architects

LANDSCAPE ARCHITECT
Jeremy Rye

STRUCTURAL ENGINEER
Conisbee

ENVIRONMENTAL SERVICES ENGINEER
Alan Clarke

PHOTOGRAPHY
Tim Crocker

Golden Lane Estate Leisure Centre, London

CITY OF LONDON CORPORATION

This project, home to the only public swimming pool in the City, is an example of retrofitting a historically significant modern building. The building's type has inherently high energy and environmental control demands, and its listed status limited the scope of fabric improvements. Nevertheless, the retrofit has succeeded in cutting energy consumption by half, making it comparable with current best practice in new building.

02

CONTEXT AND OBJECTIVES

In 1951, architects Chamberlin, Powell & Bon won the competition to design the Golden Lane housing estate. The competition, and the popular interest it generated, exemplified post-war optimism about the future and a faith in the capacity of modern ideas about architecture and planning to create coherent and well-provided neighbourhoods. The swimming pool and sports facilities were part of the estate, and with the estate they went through a long period of neglect before being listed Grade II in 1997.

The pool and tennis courts have attracted large numbers of users over the years; not only the residents of the estate, who have privileged access, but also paying customers, including people working in the many offices and businesses nearby. Refitted in the 1980s, by 2010 the leisure centre was badly in need of a comprehensive upgrade and restoration. In 2006/07, Listed Building Management Guidelines were developed by a panel of residents and stakeholders working with Avanti Architects.

In 2009 the City of London Corporation, as freeholder and estate manager, decided to modernise and expand the range of fitness options for residents and visitors, to increase income generation and reduce energy consumption. However, improvements to the spatial organisation and the environmental performance of the envelope had to respect the qualities of the original architecture and its listed status; for instance, the delicacy of the original glazing to the pool and gym had to be retained, as did the muscular barrel-vaulted club rooms, whose form was inspired by Le Corbusier's *Maisons Jaoul* (1954).

A design team led by Cartwright Pickard was selected by limited competition in March 2010. Their brief from the Corporation was to bring the building, which in the words of the project manager was 'very tired', into the 21st century. High energy performance was not a part of

01 The pool post-retrofit
02 The pool pre-retrofit

the original brief, although the client was conscious that some relatively easy wins would be possible given the very poor performance of the building.

DESIGN: GENERAL

Much of the skill of the design team went into making a modern, well-appointed leisure centre with a club feel that is attractive to customers, and creating extra space by enclosing the underused undercrofts typical of this period of architecture. The new interiors are a huge improvement, adroitly making use of the structure and spaces of the original design, the successive overlays having been stripped out. Externally, the buildings look spruced up, but only cognoscenti, alert to the thickness of glazing mullions, will be able to tell that the building envelope has been altered. In terms of energy efficiency, a very clear strategy was adopted – to make some limited fabric improvements, as permitted within the constraints of the listing, but to rely mostly on technological solutions: efficient plant and controls, a 21 kW (peak) photovoltaic (PV) array and a 50 kW air source heat pump (ASHP).

The facilities are arranged around what was once a bowling green, now tennis courts, each originally with a separate entrance. The design included the refurbishment of the swimming pool and changing rooms, conversion of the club rooms into a new gym, and reorganisation of the circulation routes to create a new single entrance and reception area. The gym was also refurbished and converted into a badminton court, and a new dance studio was added in the adjacent undercroft.

Original drawings in the City of London's archives clearly show Chamberlin, Powell & Bon's vision of a housing development as an organic part of the city. Many aspects of the original design had become eroded over the years, such as the bush-hammered concrete and brick of the vaulted structure to the club rooms, also easily identifiable in the later developments of Crescent House and the Barbican Estate nearby. These spaces

had suffered from years of severe damp penetration from above and below and had single-glazed, sliding aluminium windows that were poorly fitting and lacking in draught seals and thermal breaks. The six concrete barrel-vaulted club rooms have now been stripped of suspended ceilings, repaired and reglazed, with new uplighting onto the vaults as a main feature.

It was clear that, although originally conceived as separate buildings, the swimming pool and club rooms could be used together and would benefit from a common entrance, so that users would not have to exit and re-enter. Reorganisation to form a coherent set of spaces with clear circulation became a priority for the project. An existing plant room was removed to allow for a spacious new reception area that creates an inviting entrance. The building fabric was insulated wherever it was possible to do so without compromising architectural quality. Along the back of the space, damp and unattractive storage spaces were removed to create a new circulation 'spine', and suspended ceilings were removed to reveal pavement lights along its full length. These were repaired and restored; they now flood the space with natural light reinstating one of the key features of the original building.

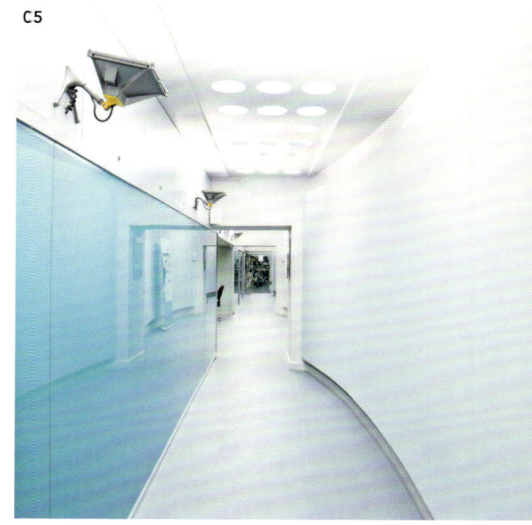

03 Overall view of the Golden Lane Estate, with the leisure centre in the middle, post-retrofit
04 Circulation space pre-retrofit
05 Circulation space post-retrofit
06 Changing room pre-retrofit
07 Changing room post-retrofit

The largely glazed facade of the swimming pool and badminton court raised the challenge of improving the performance of the building fabric while maintaining the transparency and external appearance as much as possible. Many of the single-glazed steel window frames were rusting and in need of replacement, and much of the original glazing had been replaced with opaque wired glass, not in keeping with the original architect's vision of a transparent and open structure. The solution was to re-create the original glazing using a modern double-glazed alternative. Colours and details were carefully matched to provide a high-performance envelope that is virtually identical to the original. The new windows have low-iron glass, creating visually clear double glazing.

Where new space was required for a dance studio, the design approach drew a clear distinction between new and old. The dance studio appears as a cube-like form inserted below a bridge of the existing building structure, between the swimming pool and the badminton hall. The architectural language of frameless glazing clearly differentiates this new element from the existing building. The clear glazed walls allow for maximum transparency and reduce the visual impact on the existing building; they also transform the studio into a 'shop window' that projects activity and interest to the wider estate.

08 Club room pre-retrofit
09 Club room converted to gym, post-retrofit
10 New dance studio

DESIGN: ENERGY CONSERVATION

The Golden Lane swimming pool had a number of design features that made it particularly energy inefficient, even by the standards of its time:

- uninsulated walls and roof
- single glazing
- very extensive area of glazing
- very exposed form (i.e. a high ratio of external walls and roof to usable floor area, leading to very high heat losses).

An architect designing a new swimming pool from scratch has a whole range of passive techniques to draw from: orientation, optimal glazing ratios, compact building form, intermediate buffer spaces between the warm pool hall and the cold outside, thick insulation within deep walls, triple glazing, airtight construction. However, for the architect designing the retrofitting of this existing listed building, almost none of these options were available. This meant that a rather more high-tech approach to the building systems was required.

First, any passive measures that could be introduced were used: roof insulation (but not in the walls), double glazing and improved airtightness. Then, more efficient plant and controls were adopted: demand-led ventilation with variable-speed fans controlled by CO_2 sensors (instead of on/off control), low standing loss calorifiers and LED lighting. Finally, a renewables strategy was implemented, based on a 21 kWp PV array and a 50 kW ASHP for the gym.

10

USER BEHAVIOUR

The building by its nature has a transient population, with many people passing through every day. Users have virtually no control over any of the energy-consuming elements, except perhaps the showers, which are on automatic shut-off, as is normal. There was a structured process of commissioning, training and handover, in line with the City of London Corporation's declared commitment to reducing energy use. No in-use data are available yet to assess actual energy performance.

11

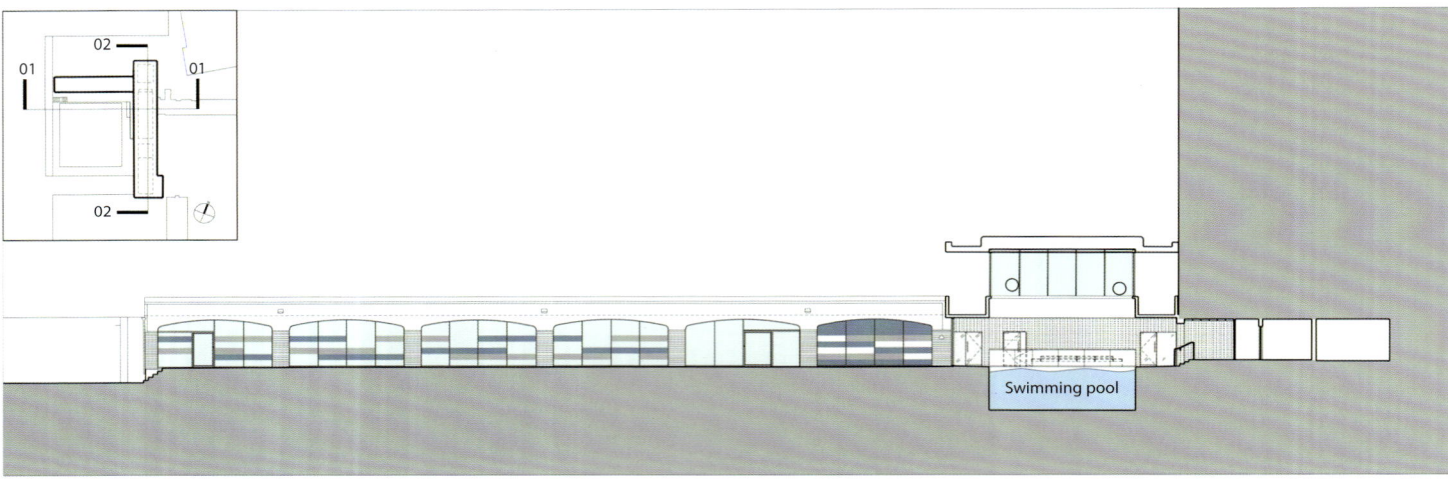

Swimming pool

12

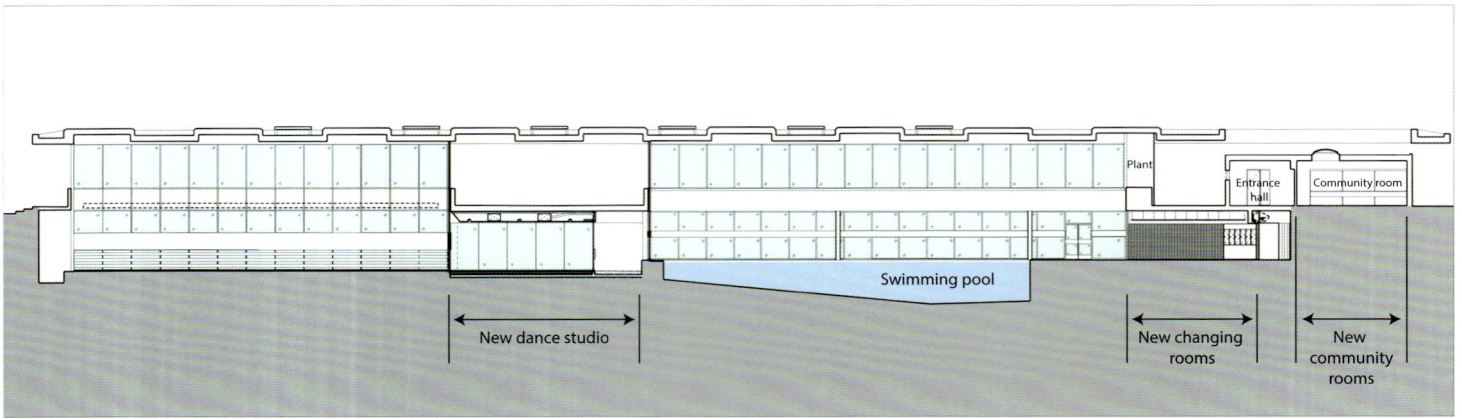

Plant

Entrance hall

Community room

Swimming pool

New dance studio

New changing rooms

New community rooms

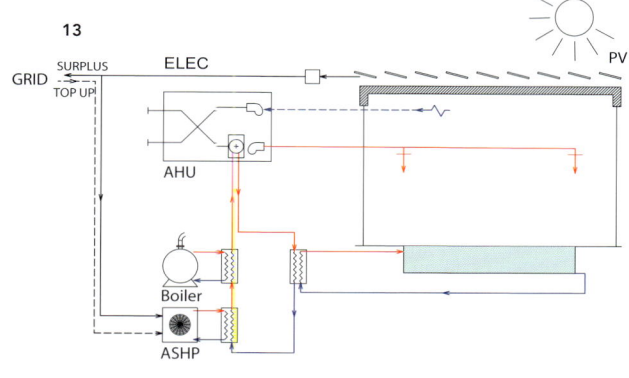

13

GRID — SURPLUS / TOP UP — ELEC

PV

AHU

Boiler

ASHP

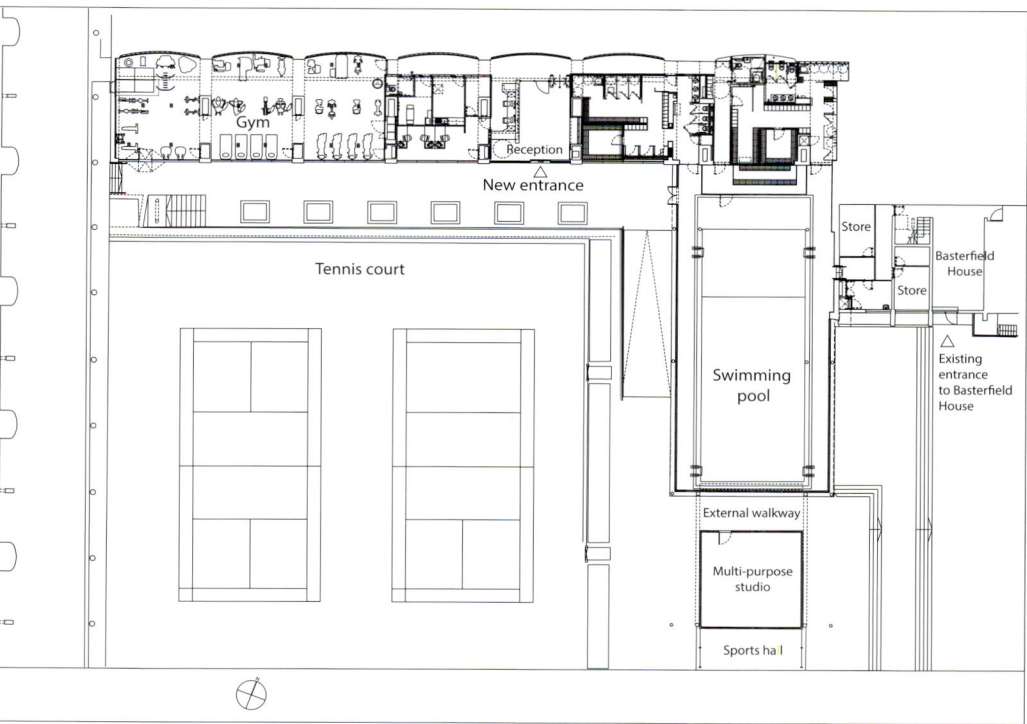

14

Gym

Reception

New entrance

Tennis court

Store

Basterfield House

Store

Swimming pool

Existing entrance to Basterfield House

External walkway

Multi-purpose studio

Sports hall

11 Section through clubroom facade
12 Section through swimming pool and sports hall
13 Schematic showing energy circuit:
AHU, air handling unit; ASHP, air source heat
pump; PV, photovoltaic panels
14 Plan post-retrofit

DATA

Energy and carbon

GOLDEN LANE LEISURE CENTRE, LONDON
Swimming pool, changing rooms, gym, sports hall and ancillary facilities

		Energy kWh/m²/yr	Carbon kg CO₂/m²/yr	Building information Area (m²)	Occupancy (m²/person)	Hours of operation
PRE-RETROFIT	Gas	1,625	325.00	1,300		
Monitored	Electricity regulated	240	120.00			
	Total regulated					
	Electricity unregulated	inc.				
	Total (est.)	1,865	445.00			
POST-RETROFIT DESIGN DATA				1,300	n/a	06:00–22:00 week
	Part L model		13.7			08:00–18:00 w/end
	Predicted energy model					
	Gas	95	19.00			
	Electricity heating	392	196.00			
	Electricity lighting	128	64.00			
	Electricity unregulated	n/a				
	Total demand	615	279.00			
	Renewables	-198	-99.00			
	Total	417	180.00			
ACTUAL DATA	Post-occupancy data	n/a				
If building in use	Gas					
	Electricity					
	Renewables					
	Total	n/a	n/a			

				Env. rating	EPC rating	
BENCHMARKS				n/a	n/a	
CONVERSION FACTORS	Gas	0.2				
kgCO₂/kWh	Electricity	0.5				

ANNUAL ENERGY USE	Electricity (kWh/m²)	Heating (gas: kWh/m²)	Heating (elec.: kWh/m²)
Typical	237	1336	
Best practice	152	573	
Golden Lane pre-retrofit	240	1625	
Golden Lane post-retrofit	128		392

ANNUAL CO₂ EMISSIONS	Electricity (kg)	Heating (kg)	Total (kg)
Typical	38,204	82,832	121,036
Best practice	24,502	35,526	60,028
Golden Lane pre-retrofit	38,688	100,750	139,438
Golden Lane post-retrofit	20,688	42,507	63,195

A building of this type is inherently energy hungry. The benchmark figures put the performance of the leisure centre post-retrofit into context, suggesting that its energy use will be comparable with best practice new build and that CO_2 emissions will have been halved. The energy figures appear better than the carbon figures, because of the shift from a predominantly gas-fired heating system to one relying much more on electricity.

If the efficiency measures alone had been implemented and gas retained as the only heating fuel, the CO_2 figure would have been 287 kgCO₂/m²/yr. By adding the PVs and using an electrical ASHP instead of gas-fired boiler plant (apart from a small amount of top-up), the emissions were brought down by 102 kgCO₂/m²/yr to give a total of 180 kgCO₂/m²/yr.

Resources and waste

If the existing swimming pool had been completely demolished and replaced with a new swimming pool which incorporated both passive design and high-tech renewables (albeit with a smaller PV array and smaller ASHP), the annual CO_2 emissions could reasonably have been reduced to around 40,000 kg, i.e. about 20,000 kg better than the refurbished scheme achieves. However, it would have had an embodied energy equivalent to 400,000–600,000 kgCO₂. Based upon a 25 year 'replace or refurbish' cycle, it's a close call whether replacement or refurbishment would have given the lowest overall carbon emissions.

Cost and time

Gross internal area:	1,300 m²
Cost per m²:	£1,769
Contract duration:	12 months
Completion:	April 2012

Procurement

The architect was appointed directly to the City of London for the design work up to and including RIBA Work Stage E of the project. This allowed the design team to work closely with the client body, helping them to fully understand and develop the brief. A number of consultations with residents of the estate were undertaken to further inform the design process. Detailed discussions were held with the City of London conservation officer and the 20th Century Society about preserving the character of the Grade II listed building. Upon completion of RIBA Work Stage E the architect was novated to the contractor. The construction contract used was the JCT Standard Building Contract 2005 edition, with contractor's designed portion. The City of London's proactive and continuing involvement ensured that quality was maintained during the construction phase of the project.

CREDITS

CLIENT
City of London

CONTRACTOR
Quinn London

OPERATOR
Fusion Lifestyle

COST CONSULTANT
Cyril Sweett

ARCHITECT
Cartwright Pickard Architects

STRUCTURAL ENGINEER
Dewhurst McFarlane

M&E ENGINEER
Synergy Consulting Engineers

PHOTOGRAPHY
Hundven-Clements Photography

15 Retrofitted facade with new glazing units

Westborough Primary School, Westcliff-on-Sea

THE GOVERNORS OF WESTBOROUGH SCHOOL

This project is part of a masterplan of works that has evolved over 17 years to keep pace with changing needs and now to significantly reduce energy use. One of several pilot projects sponsored by the UK Government's short-lived Zero Carbon Task Force, it demonstrates what can be achieved through a long-term vision and a close and sustained relationship between the architect and client. With the stakeholders – pupils, teachers, governors, site manager – all actively involved, sustainability has become an integral part of the school's ethos.

01 Assembly in the main hall post-retrofit
02 Main hall pre-retrofit

CONTEXT AND OBJECTIVES

Schools are uniquely positioned in the drive to reduce our carbon footprints. Carbon emissions from educational buildings comprise 15% of the total attributable to the UK public sector, and schools provide a fertile ground for educating young minds and, as important community hubs, raising standards in society as a whole. Building new schools makes it comparatively easy to achieve excellent energy-efficiency standards, but this is only practical for a fraction of the schools estate. Refurbishments will become more and more important in meeting the pressing need to upgrade the whole estate, even though high standards of energy efficiency are harder to achieve.

This brick-built school, dating from 1912, underwent several redevelopments before becoming grant maintained in 1995 and, following improved academic performance, an academy in 2010. Its programme of after-school activities and use by other organisations makes it an important centre for the local, culturally diverse community.

Originally there were four Edwardian school buildings at Westborough: special, infants', boys' and girls' schools. These buildings were combined into one school in the late 1950s, and in the 1970s a number of additions were built. In 1992, with the school in need of not only modernisation but also urgent repair, the governors appointed Cottrell & Vermeulen Architecture to produce a ten-year masterplan to match their vision for a well-appointed sustainable school, starting with a series of incremental improvements, such as disabled ramps, fire safety measures and general repairs. The head teacher and governors understood the value in the shared knowledge and continuity that comes from sticking with the same team and, resisting current procurement orthodoxy, which calls for competitive tendering for successive projects, have sustained the relationship to the present day.

The goal for the latest stage of the refurbishment, undertaken in three phases, was to make this ultimately a zero-carbon school, the first in the UK, and to use both the works and the finished building as a way of teaching pupils about construction and sustainability principles. To that end, the carbon-reducing features were to be made visible, not just for the children but for parents and the wider community too. The most visible element, a wind turbine, was refused planning permission following local objection. As well as reducing the school's carbon footprint, the brief was to improve the pedagogical environment, circulation and access, provide flexible communal spaces and promote outdoor learning and play, which is central to the school's philosophy.

DESIGN: GENERAL

Much of the architectural design, both generally and within the later energy-efficiency driven phases of the project, has been directed towards connecting together

03 Solar canopy from playground elevation
04 The building pre-retrofit
05 Elevation to playground post-retrofit, with solar canopy
06 Phase 2 pre-retrofit
07 Northern end of solar canopy in Phase 2, terminating in stepped play area

the separate old, cellular Edwardian buildings into a coherent whole, creating new and attractive circulation spaces and new relationships between outside and inside. To move around the school now is to experience a series of pleasant and varied places, many of which invite the school community to add their imprint and serve as showcases for their achievements.

The light and airy original classrooms arranged around shared halls have been retained and upgraded with thermal and acoustic linings, the latter much appreciated by the staff. The main school hall has

been remodelled to provide a series of flexible spaces for the school's afternoon clubs, with movable partitions allowing the spaces to be divided or opened up as needed for whole-school assemblies and out-of-hours community events. Shared practical rooms and breakout spaces for small group work have also been created in the existing building.

The most striking addition externally is a blue-roofed loggia, articulating the zigzag gable ends of the original buildings. This loggia creates an external circulation spine that is wide enough to accommodate

03

outdoor teaching, dining and playing. Supported on an expressive larch wood frame, south-facing parts of the canopy incorporate photovoltaic panels, a powerful expression of the school's low-carbon credentials, but allow dappled light through. In the playground, new 'beach huts' provide storage and further covered play spaces, and a terraced stage is used for outdoor performances.

The architect has opened the school up to its immediate urban context, binding it to the local community. A new main school entrance improves access and gives parents somewhere safe to wait, while a different new entrance allows the facilities to be used outside of school hours. As a result, the school is used much more intensively.

08 New school entrance
09 Sport in the main hall post-retrofit
10 View of main hall and side rooms

DESIGN: ENERGY CONSERVATION

The energy strategy was, in the design team's words, 'lean, mean, and green'.

The 'lean' interventions reduced energy demands from the existing building fabric through thermal insulation and airtightness. External walls and roofs were internally lined, and windows triple-glazed. The airflow through existing roof vents was controlled, while windows and doors were sealed as far as possible. Hot water and heating pipework was insulated to reduce losses and improve control.

'Mean' interventions were designed to improve energy efficiency. New boiler controls reduce unnecessary heating, while installing alternative 'regional' water heaters cuts carbon emissions from hot water by up to 50%. Heat recovery units have been fitted to optimise ventilation and heating systems. Existing fluorescent lights have been modified to use T5 lamps, with an immediate 45% energy saving, and their usage is controlled with passive infrared and daylight sensors in classrooms. Energy management of the IT infrastructure was improved, while sub-meters give a detailed record of how and where energy is consumed. The building management system monitors and controls the school's services. Carbon emissions and energy usage are displayed on an information panel in the courtyard for the benefit of students and parents. The 'green' interventions introduced rainwater harvesting and renewable energy systems, namely biomass heating and photovoltaic panels. The harvested rainwater is used for flushing toilets. Solar photovoltaic panels on the south-facing pitches of the roof structure are estimated to yield around 100 kWh/m^2/yr, while the biomass boiler is expected to more than halve annual carbon emissions from heating.

The project encountered stiff resistance to its proposal to install a 5 m-diameter wind turbine on the school site, both from neighbours and at the local planning committee. Despite officer approval, permission was refused by the committee and on appeal, a disappointing result for the design team since the turbine was an important part of the carbon-reduction mix. It is possible that greater public consultation could have made the difference, and this is highlighted as one of the lessons from the project.

USER BEHAVIOUR
AND BUILDING MANAGEMENT

A key part of the governors' vision is to maintain both the school's performance and its sustainable ethos beyond the design and construction project, and that requires changes in behaviour from everyone involved. For example, the school changed the core school day, which now starts at 8am and finishes at 1.30pm. The benefit of this is that it maximises the number of core school hours during daylight hours through the year, reducing the need for lighting. Lighting is needed for after-school clubs, but because they operate in larger groups and are contained in fewer spaces, the energy loads are limited. Activities to engage interest include participation by pupils and staff in a carbon footprint study. The efforts appear to be working, with children reportedly very aware of the project. The head teacher, Jenny Davies, who has said that the building projects on the school 'have been one of the best parts of my job' tells the story of a child who was so convinced by the low-carbon story that he became obsessed with turning off lights that were on unnecessarily. She also reports that the children who were at the school during the building works were particularly aware of the project's aims and that a higher than normal number want to go into construction and architecture.

The school's site manager, Kes Harker, is very committed to achieving the energy and sustainability aims. He has fully grasped the details of the building's operating systems, monitors energy use, and teaches the children about the environment.

11

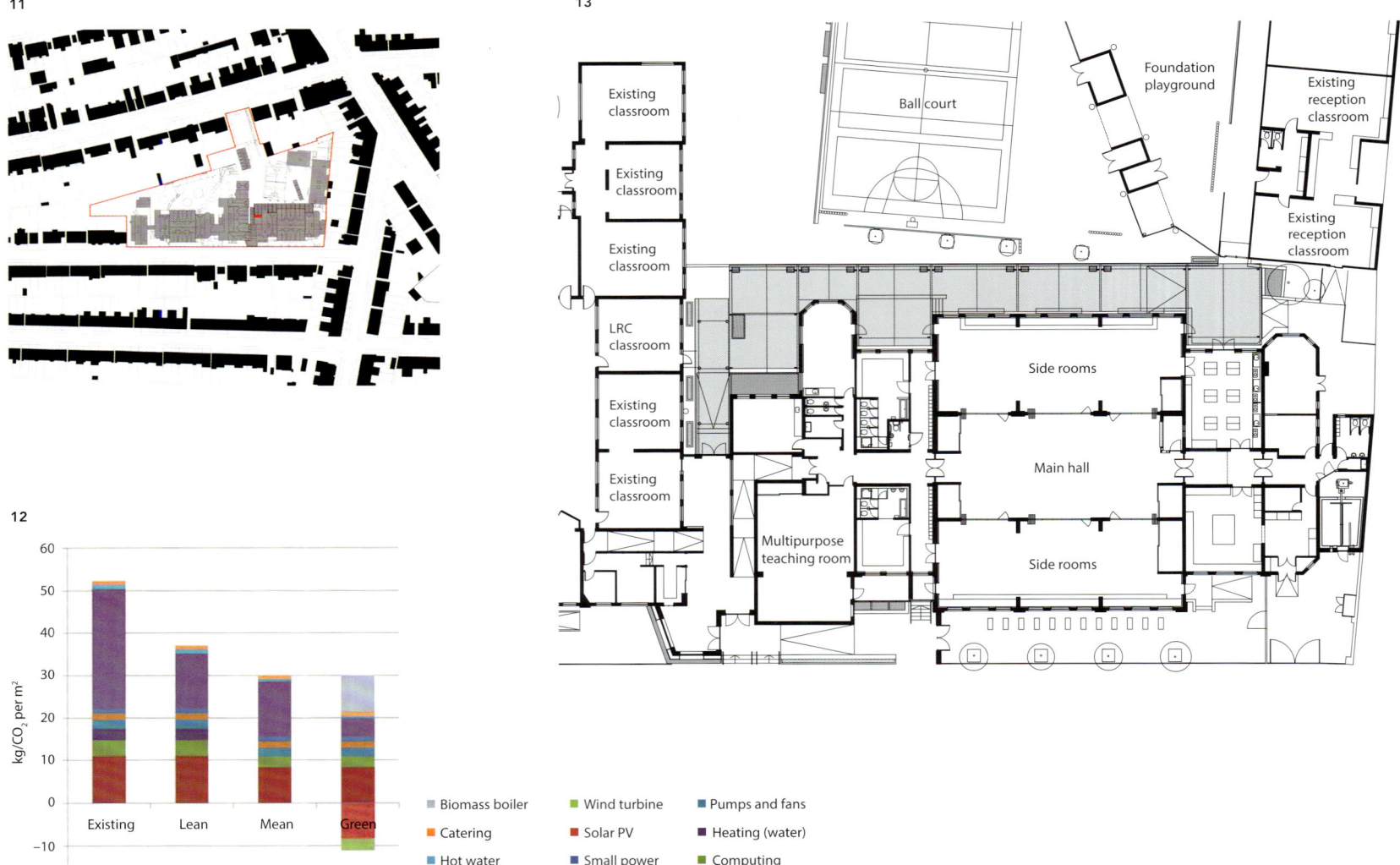

13

Existing classroom

Existing classroom

Existing classroom

LRC classroom

Existing classroom

Existing classroom

Ball court

Foundation playground

Existing reception classroom

Existing reception classroom

Side rooms

Main hall

Multipurpose teaching room

Side rooms

12

kg/CO$_2$ per m^2

60

50

40

30

20

10

0

-10

-20

Existing   Lean   Mean   Green

■ Biomass boiler   ■ Wind turbine   ■ Pumps and fans
■ Catering   ■ Solar PV   ■ Heating (water)
■ Hot water   ■ Small power   ■ Computing
■ Heating   ■ Catering   ■ Lighting

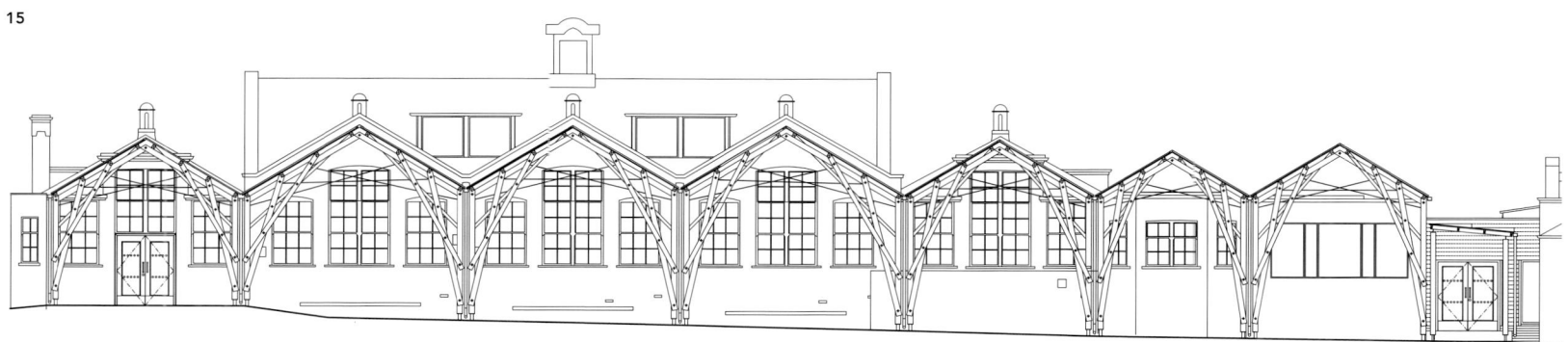

11 Phase 1 and 2 location plan

12 Annual predicted carbon emissions

13 Ground floor plan

14 Cross-section through the main hall, side spaces and canopy

15 Playground elevation showing canopy structure

DATA

Energy and carbon

WESTBOROUGH PRIMARY SCHOOL PHASE 1
Primary school

| | | Energy kWh/m²/yr | Carbon kg CO₂ / m²/yr | Building information | | |
				Area (m²)	Occupancy (m²/person)	Hours of operation
PRE-RETROFIT	Gas	152	30.4	1,023		
Monitored	Electricity regulated	55	27.5			
	Total regulated					
	Electricity unregulated	inc.				
	Total	207	57.9			
POST-RETROFIT DESIGN DATA				1,023		
	Part L model	n/a				
	Target: 60% reduction in CO₂					
	Gas		See Fig. 12			
	Electricity regulated					
	Total regulated					
	Electricity unregulated					
	Renewables					
	Total		27			
ACTUAL DATA	Post-occupancy data					
	Gas	76.4	15.3			
	Electricity regulated	56.6	28.3			
	Biomass	40.8	0.5			
	Total regulated					
	Electricity unregulated	inc.				
	Renewables (PV)	-7	-3.5			
	Total	166.8	40.6			
BENCHMARKS	ECON 73 typical practice primary school	201	47.3	EPC rating		
	ECON 73 good practice primary school	146	34.3	n/a		
CONVERSION FACTORS	Gas	0.2				
kgCO₂/kWh	Electricity	0.5				
	Biomass	0.013				

The initial target was for a 60% reduction in carbon emissions to be achieved in Phase 1, as shown in Figure 12 (previous page).

As set out in the energy table, after the implementation of Phase 1 the post-occupancy evaluation data showed a reduction of 30%. A number of reasons have been identified for the performance gap between the targeted 60% and the current operational 30% reduction:

- The kitchen, which serves the whole school, is located in the refurbished part, and accounts for a disproportionate amount of energy use if counted only within this phase. No energy-saving measures were implemented in the kitchen, therefore the pre- and post-retrofit energy use figures remain unchanged. The kitchen's energy consumption represents approximately 38% of the total electricity consumption and 17% of the total energy consumption. Typically kitchens tend to account for between 7% and 10% of energy use in schools.

- The biomass boiler stopped working for two months, which increased carbon emissions because the heating and hot water had to be serviced by the boilers only.

- Energy targets assume a standardised year for temperatures. The winter temperatures over the period of measurement were lower than average and the winter period lasted longer than the average year's.

- The number of sun hours for the annual reporting period was 33% lower than a typical year, which would have reduced the electricity generation of the photovoltaic panels.
- Two boilers serve the whole school and so it is not clear whether gas use ascribed to Phase 1 can be accurate.
- The boilers stopped working prior to the Christmas period, therefore conservative assumptions have been assumed for the energy consumption during this period.
- It is possible that the biomass boiler had not been commissioned to function at full capacity before the gas boilers fire up.

It is planned that the performance issues will be addressed as further phases are completed.

Resources and waste

Every classroom has a dry mixed recycling bin, which is collected weekly. A rainwater harvesting tank installed during the refurbishment supplies grey water to flush the toilets, and water butts connected to the new canopy are used to water the African garden and wider landscape. Compostable kitchen waste is collected and used in the school's allotment garden.

Cost and time

Gross internal floor area of refurbishment:	948 m²
Area of playground stores and canopy:	292 m²
Area served by Phase 1 M&E plant (including school kitchen):	1,023 m²
Total construction cost:	£1.35m
Cost per m² (based on total area of 1,240 m², inc. canopy):	£1,089
Start on site:	November 2009
Phase 1 completion:	February 2011
Contract duration:	14 months

Procurement

The chosen procurement method was a negotiated two-stage tender (using JCT Intermediate Building Contract with Contractor's Design 2005) to allow work to start on site before designs were fully complete. The refurbishment was phased. Phase 1 was part-funded by the DCSF's Zero Carbon Task Force and Phase 2 by the Academies Capital Maintenance Fund, while Southend Local Education Authority supported both phases of the project. Balfour Beatty also donated €95,000 (£80,000) because of the opportunities the project offered for low carbon research.

CREDITS

CLIENT
Southend-or-Sea Borough Council and the Governors of Westborough Primary School
(The project was also partially funded by the Department for Education)

MAIN CONTRACTOR
Balfour Beatty Construction Ltd

QUANTITY SURVEYOR
Stockdale

SUSTAINABILITY CONSULTANT
Buro Happold

ARCHITECT
Cottrell & Vermeulen Architecture Ltd

STRUCTURAL ENGINEER
Engineers Haskins Robinson Waters

M&E CONSULTANT
OR Consulting

APPROVED BUILDING INSPECTOR
MLM Building Control

CDM COORDINATOR
BZ Consulting

PHOTOGRAPHY
Anthony Coleman

Guy's Hospital Tower overclad, London

GUY'S AND ST THOMAS' NHS FOUNDATION TRUST

The 'brutalist' buildings of the 1960s and 1970s represent a big challenge for retrofit. At Guy's Hospital Tower the work was carried out with the tower fully and continuously occupied and the scope of the retrofit was limited to overcladding. There was no possibility of a 'deep retrofit' or of targeting the investment for maximum reduction of energy and carbon. The project was therefore conceptualised as a first step within a longer-term project to achieve low carbon emissions. The technical focus was on making the most of the limited efficiency gains possible through overclad so that this once-in-a-generation opportunity would not be wasted. At the same time, the new skin and subtly altered profile were designed to convey both continuity and contemporaneity as representations of the position and ambitions of one of the largest NHS trusts.

01 Distant view of Guy's Tower post-retrofit,
with Communications Tower in front (CGI)
02 Distant view of Guy's Tower pre-retrofit,
with Communications Tower in front

CONTEXT AND OBJECTIVES

Prominent on the London skyline, the 34-storey (143 m) Guy's Hospital Tower is among the two or three tallest hospitals in the world. Many of the building's uses have changed over its 40-year life, primarily through wards being replaced by laboratories and offices. Nevertheless, it continues to be a vital part of the facilities of the Guy's and St Thomas' NHS Foundation Trust and King's College London, containing one-sixth of the Trust's floor area. The programmatic adaptability of the building owes much to the design of the vertical circulation and services as a separate 'Communications Tower', which is distinct from the 'User Tower', with its versatile 1,200 m² floor plates which lend themselves to different functions, such as operating theatres, wards, outpatient consulting and treatment spaces, laboratories and office/admin space.

As with many commercial and civic buildings of the 1960s and 1970s, and common within the UK's healthcare estate, the tower's concrete facades had not weathered well. The Communications (or 'Comms') Tower's concrete was spalling and, to avoid big pieces crashing through the atrium below, abseilers had to periodically remove areas about to break free. The envelope was generally uninsulated, except for some cavity walls, although the windows had an early form of double glazing.

The main concern spelt out in the Trust's brief was to make the building safe and fit for purpose for the next 30 years. Competitive interviews were held to assess different teams' proposals to achieve that aim while also enhancing the appearance of this iconic building, so as to represent the modern NHS and symbolise the resurgence of this historic London neighbourhood. The Trust was committed to low-carbon principles and required sustainable and energy-efficient solutions. The winning team – with Penoyre & Prasad as architect and Arup as project manager, engineer and cost consultant – widened the definition of fitness for purpose to include a reduction in energy use and carbon emissions. They also proposed an emphatic but sympathetic transformation of the tower's appearance.

DESIGN: GENERAL

Early studies included an option to demolish and rebuild, but this was shown to be very expensive and too disruptive for the potential gain and also highly wasteful of embodied carbon.

The project scope excluded consideration of upgrading mechanical and electrical plant, confining interventions to the external envelope only. Essential to the early studies was the in-depth analysis of the existing structure to determine why the concrete was spalling and to build a picture of the building's thermal performance, including the role of the fabric, heat loss and gain, and infiltration. An understanding of occupation and use was also key to assessing the performance and effectiveness of potential solutions.

The architect positioned the overcladding as constituting the first step towards a post fossil-fuel future, a first step that should achieve the highest performance for the investment while ensuring that none of the work would need to be undone later, as and when the internal floors and associated systems are retrofitted.

Recladding options – a maximal, a minimal and a median – were developed, ranging from an external

second skin and the creation of an interstitial 'winter garden' zone to lighter-touch measures. These were evaluated in terms of cost, disruption, visual impact and energy use, assisted by extensive thermal modelling. Eventually, total surface recladding of both towers was selected by the Trust as offering the best value for the budget.

The User Tower has been provided with new floor-to-soffit curtain wall assemblies incorporating glazing and insulated panels, fitted entirely from the outside. The tower's continuous all-round service balconies, designed originally for cleaning access as well as solar shading, facilitated the installation without the need for fixed scaffolding. This solution necessitated mitigating and modelling potentially significant cold bridging, unlike the rejected option of a full external second skin with interstitial spaces. The majority of the old windows have been removed from the inside, with the work timed to suit occupants. The remaining old windows and masonry cavity spandrel walls will be removed incrementally to suit uses and budgets in this multi-tenant building. As the old walls are removed, a small but significant amount of extra floor space will be created within the rooms and spaces; each floorplate will grow by approximately 50 m², or about 4%.

03 Distant view of Guy's Tower pre-retrofit, with User Tower in front
04 Distant view of Guy's Tower post-retrofit, with User Tower in front and with the Shard to the left (CGI)

The taller Comms Tower has been highly insulated and clad in anodised aluminium panels with a stiffening geometrical fold. These have been installed using mast climbers and a service and safety platform (causeway) to protect the glass-roofed atria and areas of mechanical plant that surround the lower floors of the tower. The origami-like geometry of the mainly charcoal-grey panels creates facets at different angles, causing changing and often surprising plays of light throughout the day.

Against this folded geometry, the cantilevered form of the lecture theatre on the 32nd floor has been given flat surfaces. On the lower four floors of the User Tower's west facade – the only one that meets the

03

ground – the cladding changes to folded panels, similar to those of the Comms Tower but pale umber in colour and finely perforated to allow screened views from the spaces behind. A line of flat, pale umber cladding runs up the Comms Tower, leading to the new artwork 'crown' and marking the flues, whose terminations form the profile of the building on the London skyline.

While its concrete cladding invites use of the term 'brutalist', the tower's architecture owes much to the constructivist stream in modern architecture, with its expression of a projecting lecture theatre in the sky and the contrasting articulation of the two towers. Three possible aesthetic approaches parallel the three recladding options considered: at the minimal end, 'restoring' the tower – improving its performance while retaining its appearance – would be impractical, expensive and energy hungry as the rainscreen would have to be concrete to match the original. At the maximal end, a total enveloping of the original building was enticing, but it was considered by the Trust to be too expensive. The selected approach restores the appearance of the User Tower, with its horizontal strata brought out by the cladding and the concrete cleaning, and underscores the industrial aesthetic of the Comms Tower, sharpening its constructivist composition.

The project has also reconsidered the immediate public realm: refining the connection of the tower to the ground, improving wayfinding and improving the visitor's experience on arrival at the building. Architectural studies to consider more distant views showed how a more distinctive top would improve the

profile of the building on the London skyline. Guy's and St Thomas' Charity, well known for commissioning art, facilitated a collaboration with the internationally renowned artist Carsten Nicolai, selected through competition, to design a new crown for the building incorporating addressable LED lights which respond to movement and energy use.

DESIGN: ENERGY CONSERVATION
Pre-retrofit energy use

The tower's boiler plant, housed behind a sculptural screen by Thomas Heatherwick, had been recently supplemented by the installation of a gas-fuelled combined heat and power (CHP) plant, whose heat supplies the high-temperature hot water (HTHW) (120+ °C) ring main for the whole site, supplemented by boilers. In addition to heating, hot water and steam, the HTHW main supplies absorption chillers providing air-conditioning.

In the absence of sub-metering, energy consumption data for the tower had to be based on an area-weighted percentage of the total consumption of the whole site, calibrated by temporary meters providing two weeks' data.

The total annual energy consumption of the building was estimated to be 30.3 GWh, provided approximately two-thirds by gas and one-third by electricity. Of this, 17.9 GWh was due to internal loads, with 12.4 GWh due to solar gains, fabric gains/losses and air infiltration gains/losses. A facade refurbishment could affect only this 12.4 GWh.

05 Foot of the tower from the Great Maze Pond pre-retrofit, the only point where the facade touches the ground and connects with public space

06 Post-retrofit visualisation of the new public square on Great Maze Road

07 User Tower balconies on the left, Communications Tower on the right

08 The variations in colour on the profiled cladding is purely because of the ever-changing light

07

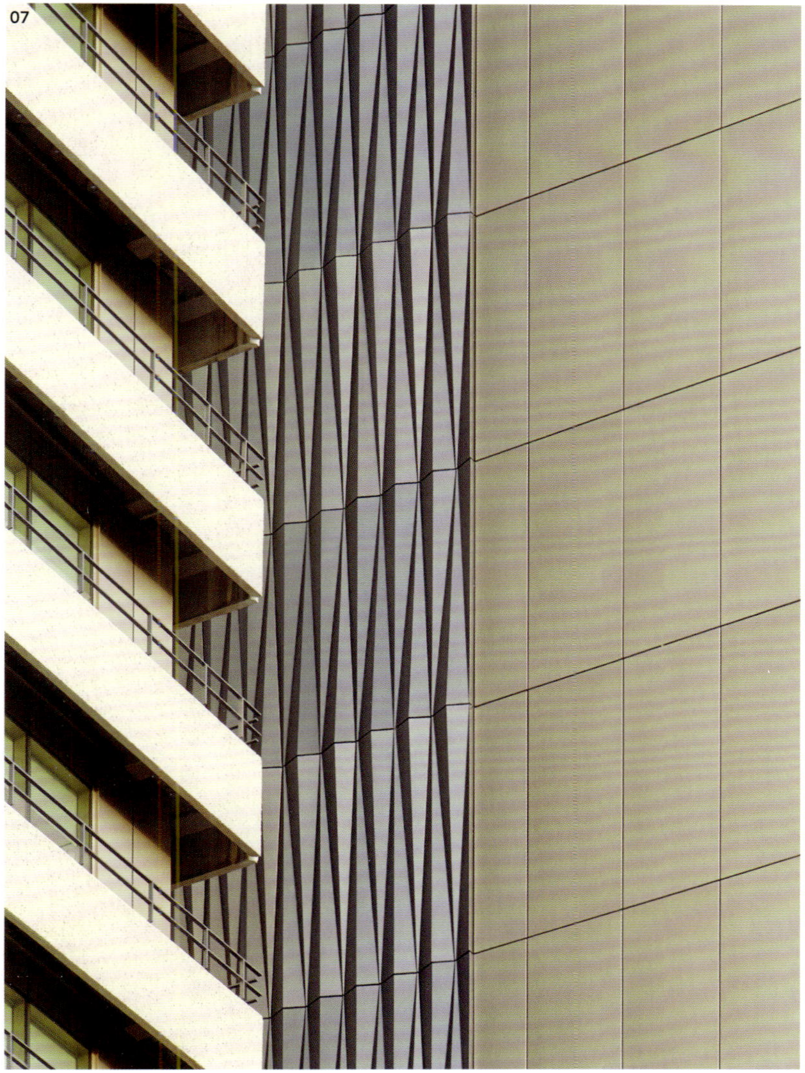

08

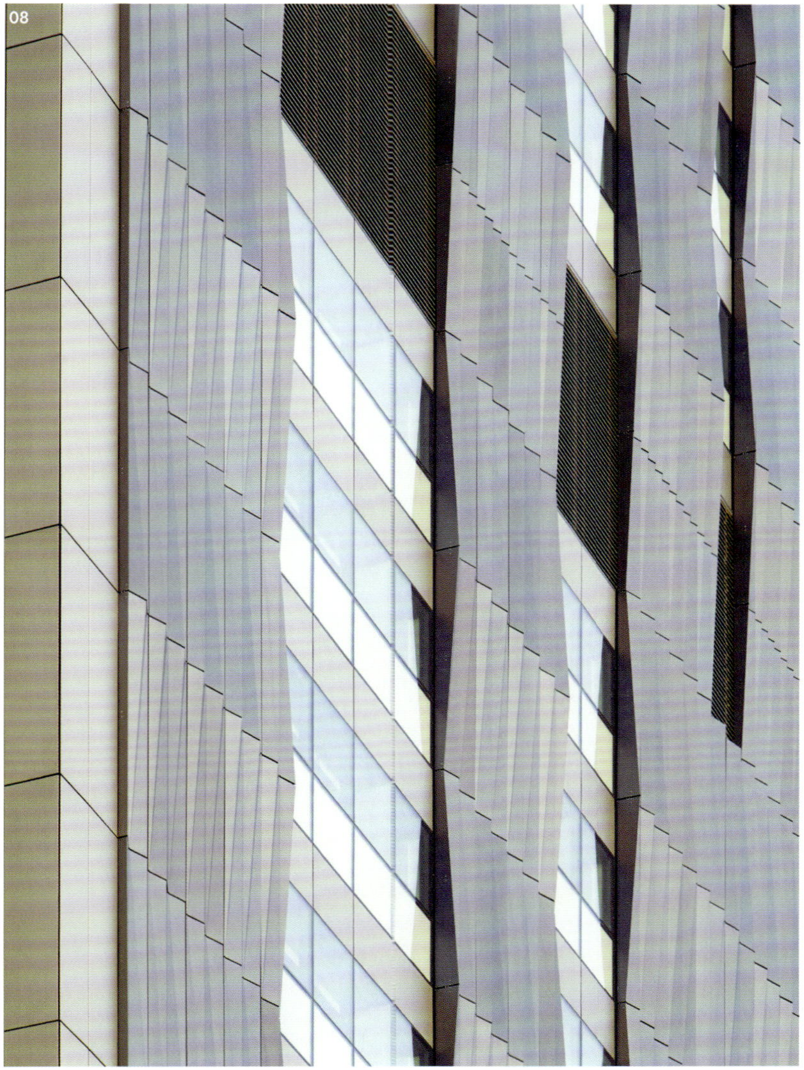

There are plant rooms on floors 3, 6, 19, 24, 29 and 30 of the tower and incrementally added plant on floors 11, 13, 14, 15 and 27. Together with numerous risers and lifts, not confined to the Comms Tower, the whole building presents an extremely complex picture of mechanical and electrical services, many parts of which have grown organically over the decades. The plant room floors are externally louvered, which means that the floor below has an uninsulated 'roof' and the floor above an uninsulated floor, both exposed to air at external temperatures. This limits the reduction of heating energy use through overcladding.

Energy performance model

A thermal model was built using the Integrated Environmental Solutions (IES) software package. The modelling initially focused on establishing a baseline, taking all prevailing conditions, such as the CHP plant, as givens, before considering relative impacts of alternative facade solutions. The majority of the heating and cooling of the building is provided via the ventilation system, but the assessment of the energy use of the tower was complicated by the presence of local split air-conditioning units, particularly in laboratories, installed incrementally by tenants.

The initial model was based on the existing building, calibrated to take into account occupant densities, lighting, equipment loads and the measured energy consumption. As well as the whole building, selected rooms were modelled to establish the impacts of differences in location, orientation, use and occupancy.

For the Comms Tower it was clear that a lightweight rainscreen over insulation, together with replacement windows, offered the most practical solution. For the User Tower, three major alternative approaches were modelled, with variants within each: replacement of windows only; addition of complete wall and window cladding over the existing facade, at the back of the balconies that run all round the tower; and total external cladding that encloses the balconies, with the creation of a 'winter garden' space. After extensive evaluation, the middle course was selected as the most effective and affordable.

Thermal performance measures

The design of the new facade incorporates generally 180 mm-thick additional insulation, significantly decreasing the U-values of the envelope, which overall averaged 4 W/m²K. Other measures comprised reduced thermal transmittance and orientation-specific low G-value glazing to reduce solar gain; low-emissivity glass where required to reduce heat loss; and improved sealing of the facade to reduce gains and losses via infiltration. While the Comms Tower has been clad in a fold-stiffened aluminium rainscreen, the User Tower uses curtain walling. Curtain walling yields higher U-values (0.8 W/m²K for the opaque areas and 1.8 W/m²K for glazing) than rainscreen (0.15 W/m²K for the opaque areas and 1.8 W/m²K for glazing), but was selected because the cladding on the User Tower had to be self-supporting.

Introduction of natural ventilation in the future

The thermal model was also used to undertake natural ventilation studies on certain key rooms, to ascertain the requirement for openable windows to support a future natural ventilation approach. The results from these studies were used to inform the design. The design of the facade ensures that low-carbon ventilation strategies, such as natural and mixed-mode ventilation, can be introduced in the future in spaces where it is possible to do so.

USER BEHAVIOUR AND BUILDING MANAGEMENT

Originally designed purely for clinical and clinical teaching accommodation, the 34 floors of the building are now used for many different purposes, with a number of different organisations and departments occupying them as tenants. As the retrofit works have no immediate impact on users' control of comfort conditions, there was no clear opportunity to engage them in the energy efficiency project. However, over time, as the plant and the systems are retrofitted, it is hoped that user behaviour, as well as building management, will play a greater part in the drive towards a low-carbon estate.

09 Perforated cladding facing public space on the Great Maze Pond
10 User Tower with new cladding and renovation of concrete balconies

09

10

11

12

13

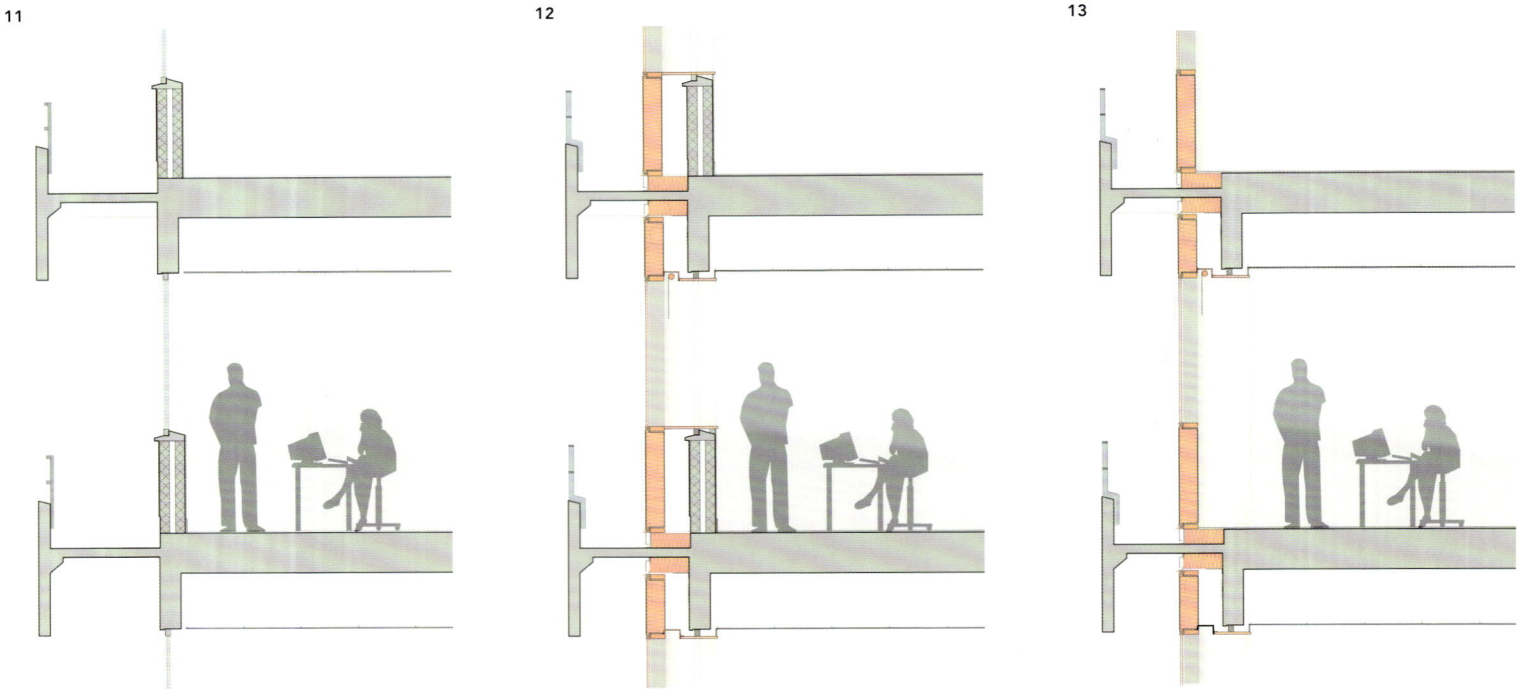

14

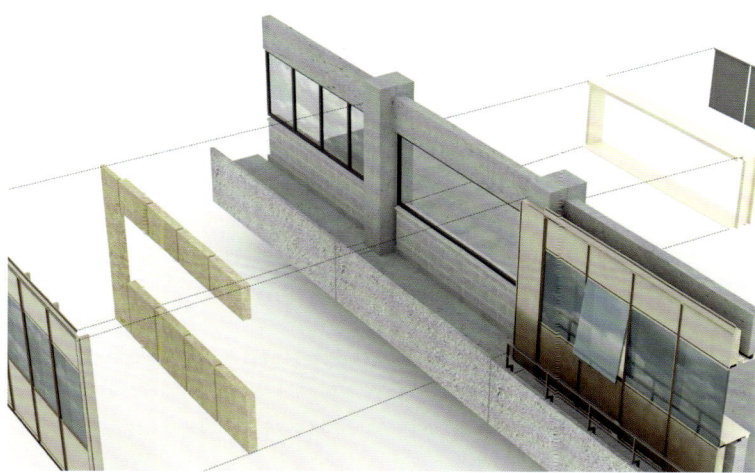

15

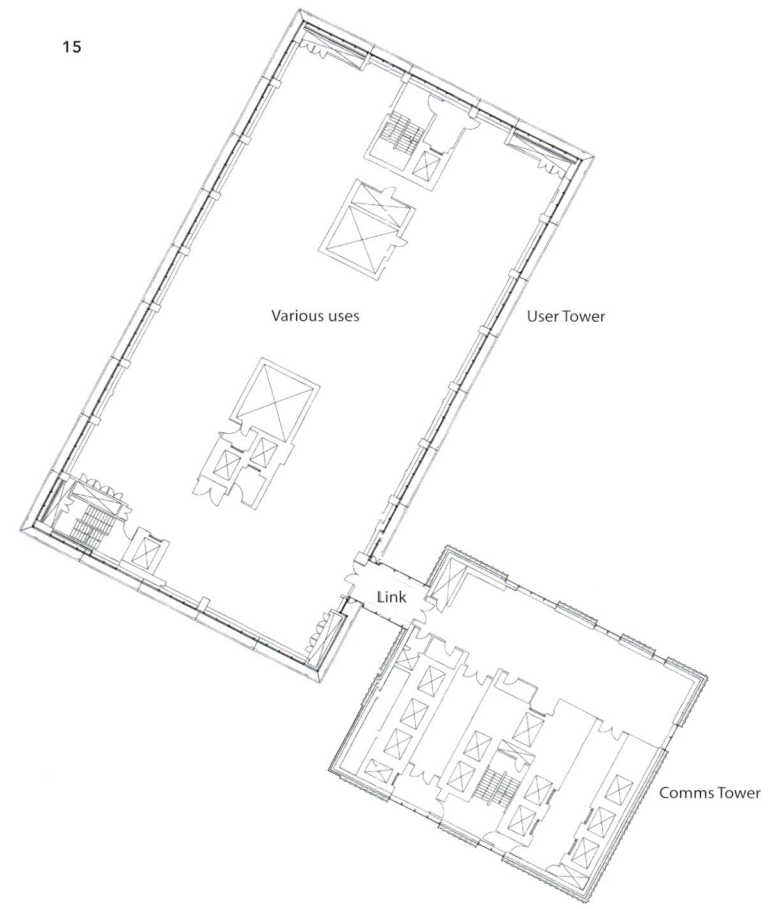

Various uses

User Tower

Link

Comms Tower

11 Facade section pre-retrofit
12 External cladding in place
13 Old windows and spandrel blockwork
   are removed incrementally to create
   additional floor area
14 Cladding assemblies for the User Tower
15 Typical floor plan post-retrofit

DATA

Energy and carbon

GUY'S TOWER
Hospital, laboratory and administrative accommodation

		Energy kWh/m²/yr	Carbon kg CO$_2$ / m²/yr	Area (m²)	Occupancy (m²/person)	Hours of operation
					Building information	
PRE-RETROFIT	Gas	358.1	71.6	56,430	10	24 hrs
Estimated	Electricity combined	178.9	89.4			
	Total	537.0	161.1			
POST-RETROFIT DESIGN DATA				56,430		
	Part L model (2006)					
	Predicted energy model					
	Gas	315.7	63.1			
	Electricity combined	180.6	90.3			
	Renewables					
	Total	496.3	153.4			
ACTUAL DATA	Post-occupancy data	n/a	n/a			
	Gas					
	Electricity regulated					
	Total regulated					
	Electricity unregulated					
	Renewables					
	Total	n/a	n/a			

		Env. rating	EPC rating
BENCHMARKS		n/a	n/a

CONVERSION FACTORS	Gas	0.2	
kgCO$_2$/kWh	Electricity	0.5	

Notes

1. Energy data based on energy model and checked against supplied data from the Trust. Note loads from the Trust were pro-rata based on what % area the Tower is of the whole estate, as energy bills were not separated by building.
2. Occupancy is an average across all floors.
3. 60% of the energy use is unaffected by the envelope. Additionally, interstitial plant room floors create uninsulated semi-external surfaces approaching the total area of the insulated cladding.

The reduction in boiler load, for both the heating system and the chillers, is estimated to be around 22%. However, the model indicates that the improved thermal attributes of the retrofitted envelope will have an effect on the cooling loads, such as to increase energy demand. Peak cooling loads are driven by the external summer temperature and solar gains, and as the new envelope will be better equipped to mitigate these, these loads will not increase. Conversely, on milder days, cooling loads are mostly driven by internal heat gains; because the better-insulated envelope inhibits dissipation of internal gains, the cooling loads will therefore increase. This is likely to increase the number of days on which cooling is required and, therefore, increase the total annual cooling load, by an estimated 26%. The building's heating load greatly exceeds the cooling load and so overall energy savings will, nevertheless, be made. If in the future a mixed-mode environmental strategy is implemented, natural ventilation should be able to deal with internal heat gains, thus reducing cooling loads.

As a result of the improvements to the facade, the energy required to offset heat gains and losses attributable to the facade (the 12.4 GWh cited above) will be reduced by an estimated 18.5%. The total building energy consumption will be reduced by around 7.6% as the high levels of electricity and hot water use is virtually unaffected by the re-clad.

Resource use and waste

A whole-life environmental impact assessment of the facade refurbishment options was carried out to assess the implications of a variety of cladding types. The environmental impact categories investigated were embodied carbon, toxicity to land, toxicity to humans, embodied water and non-renewable resource use.

Additionally, selected stages across the full life cycle were studied, including manufacturing and service life performance.

To capture the whole life cycle of each product, the additional impacts incurred in cleaning and maintaining the facade were considered. This showed that while in-service impacts are important, the initial impact of fabrication is of greater significance. The 'carbon payback' of the project using the preferred cladding option was calculated by comparing the embodied carbon in the materials against improved building thermal performance and the carbon saving this generated. The result was that the initial carbon footprint of refurbishing the facade will be recouped in just over 12 years, through the better energy efficiency of the upgraded building facade.

16 Exploded diagram of components clockwise from left: aluminium rainscreen on Comms Tower; artwork around the flues; new roof pavilion (future); solar panels over the rooftop plant (future); curtain walling system on the User Tower and link bridges; perforated aluminium cladding around base facing new public square.

Cost and time

Total floor area, including plant floors:	ca 56,000 m²
Facade area:	ca 30,000 m²
Construction cost:	£27m
Construction cost per m²:	£482 (total floor area)
	£900 (facade area)
Design team appointed:	Spring 2008
Planning consent:	Summer 2011
Start on site:	Spring 2012
Practical completion:	Spring 2014
Occupation:	The building was occupied throughout the works

Procurement

The design team was selected by competitive interview. The contractor was selected by single-stage tender and, following a validation period, a fixed price design and build contract (NEC 3) was signed. The design and project management team stayed on the client's side to manage the contract and act as supervisors to ensure the quality of the works.

CREDITS

CLIENT
Guy's and St Thomas' NHS Foundation Trust

PROJECT MANAGER
Arup

MAIN CONTRACTOR
Balfour Beatty

ARCHITECT
Penoyre & Prasad LLP

M&E, STRUCTURAL AND FACADE ENGINEERING
Arup

FACADE SPECIALIST SUBCONTRACTOR
Permasteelisa

PHOTOGRAPHY
Dennis Gilbert

16

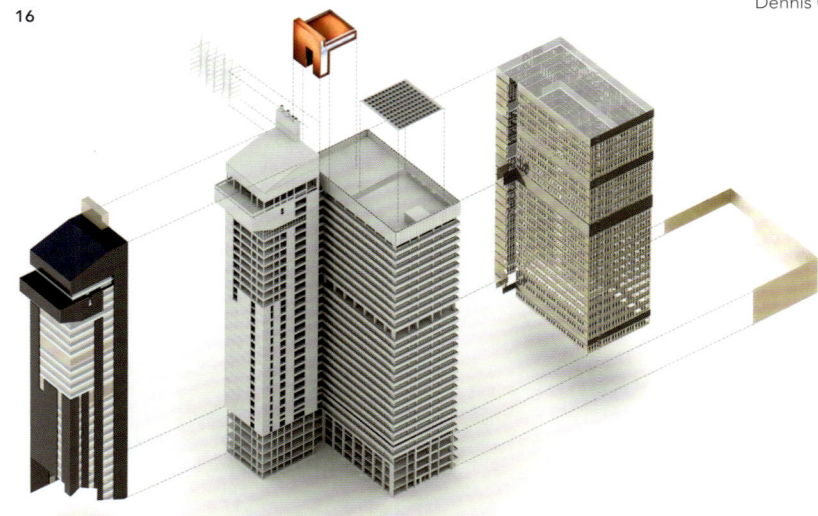

Contributors

Sunand Prasad is a Founding Partner of Penoyre & Prasad whose architectural work with a wide range of building types and with a focus on sustainable design has been widely recognised in awards and publications. Sunand was President of the RIBA, a founding commissioner of the Commission for Architecture and the Built Environment, and is involved in many initiatives to raise awareness of and find solutions for environmental issues.

Bill Bordass OBE is a scientist who took on the responsibility for environmental design and engineering at the architectural and engineering practice RMJM London. In 1984, he set up his own practice to focus on building performance in use. He is research and policy adviser to, as a well as a co-founder of, the Usable Buildings Trust, a charity that disseminates information on building performance and attempts to influence the industry, its clients and government.

Roderic Bunn FRSA, CIBSE Silver Medallist, is a principal consultant in building performance analysis at BSRIA, where he also manages the Soft Landings initiative. He has served as a Building Performance Evaluator for the Building Performance Evaluation and Invest in Innovative Refurbishment research programmes. He lectures and writes widely and is currently developing performance metrics for Soft Landings.

Richard Francis is Principal of The Monomoy Company, a strategic research and advisory firm serving clients in the built environment. He is Chair of the British Council for Offices Environmental Sustainability Group, Chair of the British Council of Shopping Centres Low Carbon Working Group, a Founding Member of the Feeling Good Foundation and an editor for Intelligent Buildings.

Professor Rajat Gupta is Director of the multi-disciplinary Oxford Institute for Sustainable Development (OISD) and the Low Carbon Building Research Group at Oxford Brookes University. His research and teaching interests lie in the areas of carbon counting, building performance evaluation, and climate change adaptation of buildings. Rajat is lead academic on numerous projects funded under the Technology Strategy Board's retrofit and performance evaluation programmes.

Matt Gregg is a Research Fellow in the Low Carbon Building Research Group of the Oxford Institute for Sustainable Development at Oxford Brookes University. He has undertaken a number of building performance evaluation studies of advanced low carbon refurbishments and has an accreditation in Leadership in Energy and Environmental Design (LEED AP).

Mark Siddall is one of the UK's leading Certified Passivhaus Designers. He is founder of low energy architectural practice LEAP, a technical advisor of the Passivhaus Trust, chair of the North East regional group of the AECB and a part time lecturer at Northumbria University.

Further Reading

LOW ENERGY RETROFIT

Swan, W. and Brown, P. (eds). (2013) *Retrofitting the Built Environment*. Wiley-Blackwell. ISBN: 978-1-118-27350-0.

Richarz, C. and Schultz, C. (2013) *Energy Efficiency Refurbishments*. Detail Green Books, Munich.

Torgal, F.P., Mistretta, M., Kaklauskas, A., Granqvist, C.G., and Cabeza, L.F. (eds). (2013) *Nearly Zero Energy Building Refurbishment: A Multidisciplinary Approach*. Spinger, London. ISBN: 978-1-4471-5522-5 (print); 978-1-4471-5523-2 (online).

Baker, N.V. (2009) *Handbook of Sustainable Refurbishment: Non-Domestic Buildings*. Earthscan/ RIBA Publishing.

Better Buildings Partnership (2010) *Low Carbon Retrofit Toolkit: A Roadmap to Success*.

Buonicore, A. (2012) Energy Efficiency Retrofit Options for the Commercial Real Estate Market. *Building Energy Performance Assessment News*.

Brack, M. (2013) Energy Retrofit of London's Commercial Building Stock. *worldcities.org*, 11 February 2013.

Moser, D., Lui, G., Wang, W. and Zhang, J. (2012) *Achieving deep energy savings in existing buildings through integrated design*. ASHRAE Transactions 118:3-10.

Liu, G., et al. (2013) *Advanced Energy Retrofit Guide Practical Ways to Improve Energy Performance: School Buildings*. Pacific Northwest National Laboratory.

UKGBC (UK Green Building Council) (2013). *Retrofit*. Available at: www.ukgbc.org/content/retrofit (accessed 1 April 2013).

BUILDING PERFORMANCE

Bordass, W. (2001) *Flying Blind – Everything You Wanted to Know About Energy Use in Commercial Buildings But Were Afraid to Ask*. Association for the Conservation of Energy.

Markus, T., Whyman, P., Morgan, J., Whitton, D. and Maver, T. (1972) *Building Performance*. Wiley, New York.

Bordass, W. (2005) *Onto The Radar: How Energy Performance Certification and Benchmarking Might Work for Nondomestic Buildings in Operation, Using Actual Energy Consumption*. Usable Buildings Trust.

Sunikka-Blanka, M. and Galvina, R. (2012) Introducing the Pre-Bound Effect: The Gap Between Performance and Actual Energy Consumption. *Building Research and Information*, June, Vol. 40(3), pp. 260–273.

Bordass, W., Cohen, R. and Field, J. (2004) *Energy Performance of Non-Domestic Buildings: Closing the Credibility Gap*. Available at www.usablebuildings. co.uk/Pages/ Unprotected/ EnPerfNDBuildings.pdf.

Leaman, A., Stevenson, F. and Bordass, W. (2010). Building Evaluation: Practice and Principles. *Building Research and Information*, Vol. 38(5), pp. 564–577.

Jones Lang Lasalle and Better Buildings Partnership (2012) *A Tale of Two Buildings: Are EPCs a True Indicator of Energy Efficiency?*.

TSB (2011). *Building Performance Evaluation, Non-Domestic Buildings: Technical Guidance*.

Zimring, C., Rashid, M. and Kampschroer, K. (2010). *Facility performance evaluation*. Available at www.wbdg.org/resources/fpe.php# (accessed 1 April 2013).

Bordass, W. and Leaman, A. (2005) Making Feedback and Post-Occupancy Evaluation Routine 1: A Portfolio of Feedback Techniques. *Building Research and Information*, Vol. 33(4), pp. 347–352.

Bordass, W., Cohen, R. and Field, J. (2004) *Energy Performance of Non-Domestic Buildings: Closing the Credibility Gap*. Available at www.usablebuildings. co.uk/Pages/Unprotected/EnPerfNDBuildings.pdf.

FINANCE AND VALUE

World Green Building Council (2013) *The Business Case for Green Building: A Review of the Costs and Benefits for Developers, Investors and Occupants.* World Green Building Council.

Muldavin, S. (2013) Beyond the Tip of the Energy Iceberg: Why Retrofits Create More Value Than You Think. *Solutions Journal*, Summer, Vol. 6(1). www.rmi.org/summer_2013_esj_beyond_the_tip_of_the_energy_iceberg_main (accessed 13 August 2013).

British Council for Offices (2012) *Change for the Good: Identifying Opportunities from Obsolescence.*

Francis, R. and Wright, W. (2012) *Counting Carbon, Counting Costs: Achieving Performance in Retail Fit Outs.* British Council of Shopping Centres.

Mills, E. (2009) *Building Commissioning: A Golden Opportunity for Reducing Energy Costs and Greenhouse Gas Emissions.* Lawrence Berkeley National Laboratory.

TECHNIQUES & TOOLS

Bordass, W., Bunn, R., Leaman, A. and Wray, M. (2009) BSRIA BG4/2009 *The Soft Landings Framework – For Better Briefing, Design, Handover and Building Performance In-Use.* BSRIA & The Usable Buildings Trust. ISBN 078 0 86022 7.

Bunn, R. (2013) BSRIA BG 45/2013 *How to Procure Soft Landings – Specifications and Supporting Guidance for Clients, Consultant and Contractors.* ISBN 978 0 86022 719 9.

CarbonBuzz. *The performance gap.* Available at www.carbonbuzz.org/.

Feist, W. (2012) *Criteria for Non-Residential Passive House Buildings.* Available at http://passiv.de/downloads/03_certfication_criteria_nonresidential_en.pdf (accessed 18 June 2013).

Feist, W. (2012) *Certification Criteria for Energy Retrofits with Passive House Components.* Available at http://passiv.de/downloads/03_enerphit_criteria_en.pdf (accessed 18 June 2013).

BREEAM. (2012) *BREEAM In-use.* Available at www.breeam.org/page.jsp?id=373 (accessed 1 April 2013).

IMAGE CREDITS